Willa Zakin

About the Author

CHRISTOPHER HALLOWELL, a professor of English and journalism at Baruch College, City University of New York, has a special interest in cultural science and environmental journalism. He is the author of several books, including *People of the Bayou*, a narrative account of Cajuns who live off the land in South Louisiana. He is also coauthor and editor of *Listening to Earth*, a collection of writing that focuses on social and cultural issues surrounding environmental conflicts, and *Green Perspectives: Thinking and Writing About Nature and the Environment*, an historically organized collection of writing that traces the course of attitudes toward the environment in this country throughout the past century. He has been on the staff of numerous magazines as an environmental or science editor and has contributed articles to *Time*, the *New York Times Magazine*, *Christian Science Monitor*, *Audubon*, *Geo Magazine*, *Natural History*, and *The American Scholar*. He has reported from Peru, Panama, the South Pacific, and East Africa, as well as from various regions of this country. Professor Hallowell, a former director of undergraduate journalism at Baruch College, is a graduate of Harvard and received a master of science degree in journalism from Columbia. He lives in Brooklyn, New York.

Holding Back the Sea

Holding Back the Sea

The Struggle
on the Gulf Coast
to Save America

Christopher Hallowell

HARPER  PERENNIAL

NEW YORK ● LONDON ● TORONTO ● SYDNEY

HARPER ● PERENNIAL

FIRST HARPER PERENNIAL EDITION PUBLISHED 2005.

Book design by Nancy B. Field
Endpaper map by Karen A. Westphal

The Library of Congress has catalogued the hardcover edition as follows:

Hallowell, Christopher.
 Holding back the sea : the struggle for America's natural legacy on
the Gulf Coast / Christopher Hallowell.—1st ed.
    p. cm.
Includes bibliographical references and index.
ISBN 0-06-019446-4 (alk. paper)
1. Wetlands—Louisiana. 2. Wetland ecology—Louisiana. 3. Louisiana—
Environmental conditions. I. Title.
QH87.3 .H35   2001
333.91'8'09763—dc21   00-143910

ISBN-10: 0-06-112424-9 (pbk.)
ISBN-13: 978-0-06-112424-2 (pbk.)

05 06 07 08 09 ❖/RRD 10 9 8 7 6 5 4 3 2 1

To the people of South Louisiana, who have struggled so long to save their coast, may the irony of Hurricane Katrina be that she inspired others to join the effort.

Acknowledgments

Knowing what and who to acknowledge is difficult. There are so many of each. Why I have been fascinated by the uncertainty of edges ever since I can remember—and edges are endemic to wetlands where land seems to melt into water—I am not sure. It might have to do with childhood summers on Cape Cod and the always-mysterious tidal pool at the Wing's Neck Beach. It might have to do with growing up on a farm and having the freedom to explore its brooks and ponds.

A series of happenstances occurred without which this book could never have been written. I happened to go to New Orleans years ago and as the plane approached the airport, I looked down upon a wetlands world I never knew existed. I have been mesmerized by it ever since. At some point I met Cathy Norman, of New Orleans, through someone I was interviewing for a magazine article. Cathy was an invaluable resource, not least for introducing me to her husband, Shea Penland. Cathy and Shea's concern about and familiarity with the Louisiana wetlands was always an inspiration. Their willingness to share their time and knowledge goaded me on. Shea introduced me to many of his coresearchers to whom I am also indebted: Denise Reed, Gregory B. Miller, S. Jeffress Williams, and Gordon Helm, to name but a few.

Many other people were exceedingly generous with their time and knowledge, a testament to their concern for the plight of the great wetlands that Louisiana harbors and that is so vital to the rest of the country. Among them is Donald W. Davis, who kindly answered my barrage of questions about the historical development of Louisiana's marsh. I also acknowledge former Louisiana state senator Dr. Michael Robichaux for his caring about the people of South Louisiana. I am indebted as well to William H. Burton, Vincent F. Cottone, and Frederick Palmer, oilmen eager to explain arcane aspects of their industry. I also want to thank Cythnia Sartou and Doug Daigle, who willingly shared their opinions about human impacts on the Gulf of Mexico. And thanks to Richard Boe, an Army Corps of Engineers biologist, who introduced me to the excitement of fishing for speckled trout and redfish.

I am grateful to Sally Stassi of the Historic New Orleans Collection for so enthusiastically searching out old maps of Louisiana and New Orleans.

Another special thanks goes to Karen A. Westphal, who worked a miracle in a very short time and created a fine map for this book. Thanks also to Philip Gould, a tireless and marvelously creative photographer, to Sandra Russell Clark and Evert Witte for their ongoing encouragement, and to Walter Levy for his determined reading and probing questioning of earlier stages of this book. Thanks to Gene Moncrief, too, for her excellent presentation coaching.

Acknowledgment is also given to the City University of New York Research Foundation for partial funding of this project. I also want to thank the University of New Orleans for its support.

Thanks to my editor, Hugh Van Dusen, who had the wisdom to support this book despite the skepticism of others, and to my agent, John Thornton, who is always encouraging.

Lastly and mostly, thanks and love to Willa Zakin for her generosity of spirit and her appreciation of the people and places of South Louisiana.

In human history, we have learned (I hope) that the conqueror role is eventually self-defeating. Why? Because it is implicit in such a role that the conqueror knows, *ex cathedra,* just what makes the community tick, and just what and who is valuable, and what and who is worthless, in community life. It always turns out that he knows neither, and this is why his conquests eventually defeat themselves.

—Aldo Leopold
from *A Sand County Almanac*

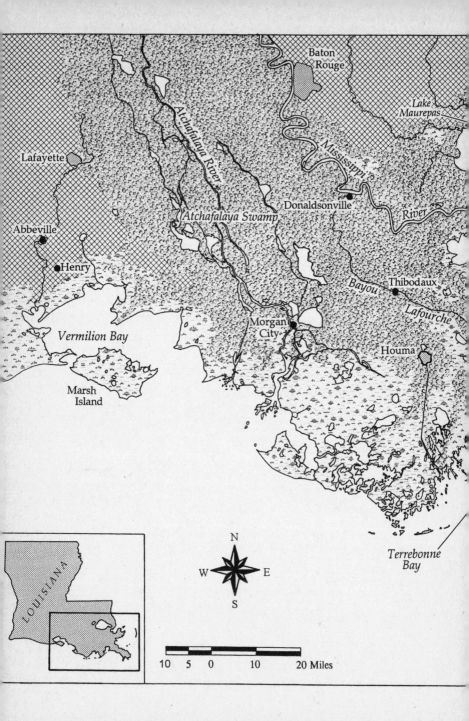

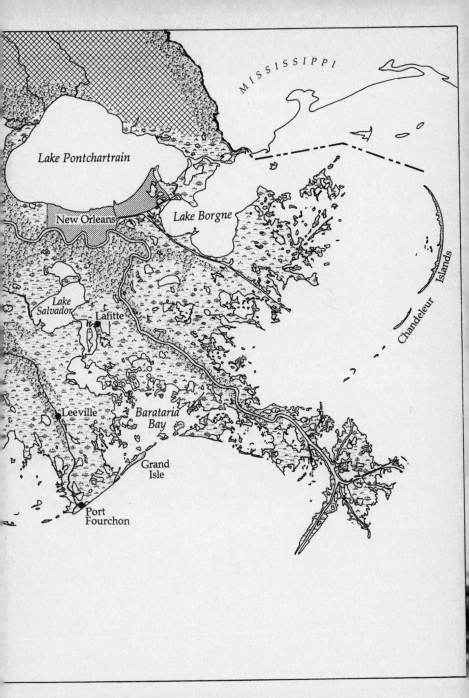

MISSISSIPPI

Lake Pontchartrain

New Orleans

Lake Borgne

Chandeleur Islands

Lake Salvador

Lafitte

Leeville

Barataria Bay

Grand Isle

Port Fourchon

Contents

Introduction

This book, first published in 2001, predicted the inevitability and the consequences of a hurricane such as Katrina. I didn't have to stare into a crystal ball. It was simply a matter of putting the pieces together. The vast watery region of South Louisiana was a place I had first visited and come to love in the late 1970s when I began researching an earlier book, *People of the Bayou*. I came to realize that the lives of the mostly Cajun bayou dwellers, people who had welcomed me into their homes and their lives, and whom I had come to know as I accompanied them in their seasonal cycles of oystering, shrimping, crawfishing, crabbing, fur trapping, and rice planting, were doomed to change. These people lived and made their precarious livelihood and complained often that the wetlands were sinking and fragmenting beneath their feet and camps. Something was going wrong with their environment. That realization was for me the beginning of Hurricane Katrina.

Over the years the possibility of a Katrina-like storm loomed in my mind. In 1998 I returned to Louisiana to research and write *Holding Back the Sea*. I soon discovered that those changes to the environment were relentlessly transforming Cajun livelihoods and culture but also threatened to destroy the entire Gulf coast on a far larger scale. Many of the causes were manmade—based on special interests and short-term thinking. The stage was set for a huge hurricane—a Katrina—to wreak destruction on the great city of New Orleans and the sweeping wetlands of South Louisiana. No crystal ball. Scientists, engineers, oil- and gasmen, and shrimpers and oystermen all knew that the Big One was coming and that it was only a matter of time.

People showed me their bayou homes standing in water; coastal roads that flooded over with increasing frequency; houses in East New Orleans that tilted on cracking slabs, and front steps that were pitched out of kilter by the quickly subsiding ground beneath them; and the Gulf's saline waters creeping up bayous and beginning to contaminate water wells far inland.

The wetlands' march toward extinction seemed relentless—an area the size of Manhattan—twenty-five square miles, or sixteen thousand acres per year—a football field every fifteen minutes; people who worried about such things employed a variety of descriptors. When New Orleans was founded in the early eighteenth century, five million acres of wetlands skirted what is now the Louisiana coast, a nearly three-hundred-mile-wide swath penetrating almost one hundred miles inland in some places. Two million acres of this huge expanse have disappeared, some replaced by communities, cities, parking lots, and Walmarts, but much of the fragile tapestry has simply unraveled into the Gulf of Mexico.

The dissembling picked up speed as protective levees along the Mississippi grew higher and higher, initially at the hands of the French, then the Spanish, then the French again, and finally by the Army Corps of Engineers, charged in the wake of the massive flood of 1927 to protect the city and to keep the river navigable. The leveeing starved the wetlands of nutrients and the replenishing sediment from the Mississippi's annual rites of flooding. The land began sinking, compacting, and fragmenting, enabling the Gulf to encroach northward.

The process accelerated in the early decades of the twentieth century when the nascent oil industry began dredging canals through the wetlands to bring in drilling equipment and lay pipe. It's an incredible sight to look down from a helicopter or plane and witness what they did: canals slice through the wetlands, crisscrossing one another in a stark weave contrary to nature's design. The viewer can only gawk at the apparent desperation of humans so bent on nature's riches, as to slash through the soft land like some berserk butcher.

By 1998 I witnessed increasing worry during many tedious meetings where bureaucrats and technocrats, big landowners, shrimpers and oystermen, and local academics and politicians argued for hours in stuffy rooms about how to fix the eroding wetlands, the subsiding of New Orleans, the encroaching Gulf, the voracious nutria chewing away at the marsh's roots leaving slimy mud flats that turn into a pond in the next flood, and then a lake, and then join the Gulf forever. Many solutions were offered: plant the barrier islands offshore with vegetation and build dunes on their sand backs; flush sediment into the wetlands from the Mississippi through expensive control structures inserted into the river's levees; pump sediment from the bottom of the Mississippi; pump sediment from offshore; fill in canals, fabricate jetties offshore, sink old barges offshore, and fill them with rocks to create instant barrier islets.

No ideas were sufficient to counter the destruction of South Louisiana that human beings had begun and that nature was in the process of completing.

As concern over the vanishing wetlands and the sinking coast grew, and plans were offered to alleviate the situation, voluble denial set in alongside the worry. Eyes began to harden as people dug in their heels, holding onto their special interests with increasing intensity. Landowners didn't want the state or the Army Corps of Engineers messing with their land; oystermen didn't want their oysters exposed to Mississippi River water by some cockamamy freshwater diversion project that would bury their beds in sediment. Oil company representatives claimed they needed those canals that had been dredged through the wetlands so many years before, even though the wells they led to had long gone dry; the transportation industry wanted the navigation canals to New Orleans and Houma to remain open, though they had tripled in width due to erosion of their soft banks as boat wakes and storm surges chewed up their banks year after year and storm after storm.

South Louisiana, it occurred to me, might be doomed just through the endless arguments over how to save it.

No one could agree on a plan—a grand plan. But eventual compromises of sorts worked into all of the discussions. South Louisianans are friendly people, promoted by a fickle environment of storms and floods where cooperation in food-gathering and harvesting have always paid off in the end. The bickering died away and small-scale agreements were struck—an experiment with sediment pumping here, a barrier island reinforced there, a stone jetty somewhere else—inconsequential actions in the face of the huge loss of wetlands, actions that would cost little and compromise no one. The federal government would pay the bill along with some state assistance. Those who attended such meetings parted on back-patting terms, though they may have totally disagreed with one another across the table.

One of the first things I did when I returned to Louisiana in 1998 was to head for Thibodeaux on Bayou Lafourche where former Representative Billy Tauzin was putting on a day-long conference at Nicholls State University on the dangers of sinking marsh, rising sea levels, and the prediction of more frequent hurricanes posed to South Louisiana. The conference took place a couple of weeks after Hurricane Georges had just missed New Orleans, turning east at the last minute and slapping Mississippi a relatively mild wallop. It was a close call; fear was in the air. All sorts of people attended the conference, from worried fishermen to strangely blasé state officials to a group of elementary school kids, who were prompted by their teacher to read aloud letters they had written to President Bill Clinton asking him to help save their wetlands. The kids were cute and their letters were cute and they set the tone for the day's event. The audience chuckled and turned mushy. An engineer told a funny story about how the rising waters would force everyone to buy taller boots; the state people tried to defend their lack of action; and when some scientists talked seriously about the vanishing wetlands and how they were so valuable as a buffer against hurricane storm surges, the audience twitched with embarrassment because they knew what was happening to their state and there didn't seem to be anything anyone could do. Talk of the Big One was met with quick grimaces because people realized how awful it would be, but what could be done? And maybe New Orleans, anyway, was blessed because it had not been badly hit since Betsy, in 1965. After all, so many hurricanes had headed right for the city and then veered east or west at the last minute. Denial still ruled.

Governor Mike Foster drove down from Baton Rouge for the lunch. A lot of people thought that was a bad idea because he had not expressed a whit of interest in saving the wetlands. They were right. His speech after lunch reflected a huge lack of interest in the problem. At the podium his eyes glazed over and he went on automatic pilot, telling stories about how much he loved wetlands and duck hunting. He said virtually nothing about the danger facing Louisiana.

People went away from that conference confused, paralyzed, and full of self-blame. They didn't turn their feelings against the government, nor the Army Corps of Engineers, nor the oil industry. They blamed themselves for not doing anything to stop the land loss. Out of that frustration, they began to think in some new ways. Evidently, Governor Foster began to think differently as well, and before long, pivoted 180 degrees and launched a fight for the wetlands. It was the beginning of a transformation in the way the people of South Louisiana—not the business people or the oil people or the state people, but the People—began to see that it was up to them to save their part of the state.

Coming from many walks of South Louisiana life, they took to quietly protesting the ruination of their coast, driven by the realization that no matter who they were—shrimper, truck driver, tourist guide, roustabout, welder, or mayor—so much of the economy, their economy, was based on the health of the wetlands. Their lives and livelihoods were at stake to say nothing of their way of life from work to music to food. Reluctant at first, business people, big landowners, and oil executives joined the group. The result was the publication of a 161-page document entitled *Coast 2050: Toward a Sustainable Coastal Louisiana*, the first unified plan to save the wetlands, New Orleans, and bayou communities, through wiser implementation of wetlands building and restoring techniques.

It would cost fourteen billion dollars. The federal government said "no"—no to footing the bill for the preservation of one of the most important contributors to this country's economy. Both the president and the Congress refused to pay more than a fraction of the requested amount. Louisiana was left on its own to preserve a national treasure. Italy, Great Britain, the Netherlands, and Japan have put billions into preserving their coast lines and national treasures. Our government has not and now we must all pay a huge price.

Shortly after *Holding Back the Sea* was first published, Governor Mike Foster, spurred by his recent transformation from a wetlands lover to a wetlands saver, held a big conference in Baton Rouge, a one-day affair to publicize the matter. He invited me to give a keynote address before four hundred or so people who attended, which I did with pride. In his introduction of me, Foster said he and George W. Bush were friends and he had given the president a copy of the book. Bush had telephoned Foster and told him he had it on his bedside table in Crawford, Texas, and planned to read it. The governor heard nothing more from the president about the book, the wetlands, saving Louisiana, and certainly nothing about the role of the oil industry's role in destabilizing the already precarious wetlands environment. (This was before 9/11.)

With Mike Foster's blessing, the backing of Jack Caldwell, then Louisiana' s secretary of the Department of Natural Resources, who had given a copy to every member of the Louisiana legislature, and with a fine review in the *New Orleans Times-Picayune*, the book seemed to be well positioned.

Outside of Louisiana, however, it had few readers despite a plethora of strong reviews and a string of radio and television appearances. I began to identify a pattern of ignorance about the reality of Louisiana and the Gulf coast. Many people who attended my presentations and readings had visited New Orleans, appreciated its food and music and history, but had no sense of New Orleans, the place, and certainly little awareness of the immensity and fragility of its surrounding wetlands. New Orleans was a fixed point in their minds—a place to go to do certain things, someplace where you could expect to see funky people, and let yourself soak up its legends of loose behavior and above-ground graves. Few people envisioned New Orleans as a bowl waiting to be filled with water, a city that was 60 percent below sea level. It was even more foreign for people to grasp the significance of the Gulf coast to the rest of the country—that it nurtured and produced 25 to 35 percent of our total seafood catch, that 25 percent of our oil and gas is pumped out of the wetlands and offshore, and crosses the vanishing marshes in a maze of pipelines on the way to refineries along the Mississippi River. The idea that one of the poorest states in the Union with a long history of political corruption could be crucial to the welfare of the rest of the country did not fit in with preconceived notions garnered from crass Cajun jokes, Mardi Gras frenzies, Bourbon Street binges, and *laissez les bons temps rouler.*

Then came Katrina.

This book is about the South Louisiana that faced Katrina, a region so environmentally mutilated that the hurricane's winds and waves had an easy time drowning New Orleans and wiping away the surface of Plaquemine and St. Bernard Parishes, as well as the Mississippi and Alabama coasts. How did the mutilation occur? How could three million acres of wetlands—40 percent of this country's coastal wetlands—be so ignored? Why wasn't the levee system protecting New Orleans strengthened in the knowledge that hurricanes now were almost certainly more violent? Why were evacuation plans so vacuous? How could this country not realize that one-quarter of our energy was piped across the vanishing wetlands? And ultimately, how could our federal government have allowed this tragedy to occur? This book is the inside story of how these misfortunes came to pass.

The book goes beyond Katrina, beyond the bayous and marshes of South Louisiana, beyond the Gulf coast and New Orleans. While nothing could have stopped Katrina's force, preparation could have saved many lives and much of what is our *shared* heritage—a heritage that the unique and wonderful region of South Louisiana fostered.

The story of the plight of the Gulf's vast but fragmenting wetlands is full of lessons. One of the most important is the respect for nature. Another is that for all the bravado with which we often disregard nature, we inevitably come to the realization that we can do *no* better than to imitate her. Our hubris is humbled. In this eventual awakening lies the only hope for the salvation of this crucial, delicate, and rich environment. Its wholesale abuse cleared the way for Katrina's relentless destruction. The manipulation of this environment has been so extreme and so damaging as to threaten economies on a vast scale. The loss of livelihoods, of the way millions of people have lived for generations, must also bring the acknowledgment that maybe nature is right and its dictates must be lived with rather than ignored or conquered.

Holding Back the Sea is a wake up call to America—Americans of all regions and walks of life—to special interest lobbyists working their narrow deals as they pace the halls of government. It is also a plea to our government. We are no longer protected in our urban high-rises, air-conditioned rooms, suburban homes, yards, malls, and country clubs. We are no longer protected in our country cottages and farms—set in rolling hills, mountains, and flat prairies. We are no longer protected in the Congress and the White House. We are all connected to what happened in Louisiana and along the Gulf coast. The nature and power of our interconnectedness will continue to be revealed long after Katrina fades from our immediate attention.

Since the original publication of this book, some of the many people who fill its pages have changed jobs and paths to be replaced by other personalities. Major changes include the following: Governor Mike Foster did not seek reelection in 2002 and Kathleen Babineaux Blanco is now governor; Senator John Breaux has retired, as has Representative Billy Tauzin; almost the entire cast of the Louisiana Department of Natural Resources has been replaced; Frank Hijuelos is no longer with the Office of Emergency Preparedness for the city of New Orleans; the Army Corps of Engineers officials mentioned have been replaced by others in the frequent rotations that are endemic to that institution. Otherwise, most of the people mentioned in the original edition are still in the same place physically. Yet they have undergone a sea change, for Katrina and her aftermath have had a profound and enduring impact on the physical and emotional landscape of South Louisiana and beyond.

Holding Back the Sea

Chapter One

Jim Daisy's Legacy

Now Jim's dead. His nephew, Jeff, tells me this, leaning over the balcony railing of his big square house up on stilts on Bayou du Large, looking down at me, and not looking too friendly. The house rests on a forest of stilts. Between their trunks I can see a half-submerged lawn to the rear of the house. It is littered with the sunken wrecks of toys, as the surrounding marsh inexorably sinks and the Gulf of Mexico creeps closer. Beyond the lawn, skeletons of drowned live oaks and cypress predict the future, their naked arms raised to the beating sun

"Uncle James, it must have been 'bout three years ago that he passed on; he mus' a been 'bout sixty-eight years old." Jeff bit at his thick red beard, as if wanting to grab back the words he had just spoken. "Why you askin' ?" He did not know who I was, only that I was not local. I did not look like a tourist, and probably not like an oyster dealer, either.

The only other strangers to come down the bayou are selling religion or some kind of sustenance for the emotions or the heart. To the five hundred or so residents of Bayou du Large, and most other South Louisiana bayou towns, such people are nettling, an irritant reminding them of the precariousness of their lives as the contemporary world, ordinarily visible only on their TV sets, now encroaches with greater frequency, generally from the north. Then, from the south come the Gulf's waters. Both are on the verge of sweeping away the lives of these people.

It had been ten years since I had driven down the oyster shell-lined road along the bayou. Both the road and the bayou begin on the Gulf side of Houma, dividing flat land cleared for sugarcane, both traveling straight south. Then they undulate sinuously across the land, and the road—which came long after the bayou—is hard put to keep up with the bayou's sensuality, pitched too hard into its curves, sometimes seeming to stumble right into its slow water.

The fields give way to palmetto scrub as the bayou winds toward the Gulf and the fishermen's houses and boats begin to cluster its edges like oysters on a reef. Things are messy down here. Rusting engines, smashed up cars, metal struts, tanks, all sorts of stuff that are hard to figure out, line the road on the bayou side.

Abandoned shrimpers and luggers lurch up on the banks like they plowed right into the mud after a bad night. Their carcasses rot fast, turning muddy and green. Down here, people don't get nostalgic about boats. Their lives depend on them; when the stem keels and ribs begin to rot, they're through—stripped of their equipment and set out to die. Planking is another matter; it can be replaced. On the other side of the bayou, naked Lafitte skiffs, with a woman's curves in the bows and a big square tail, lie upside down with ribs bare waiting for new cypresses. During winters, older men up and down the bayou caress designs for their boats with pencils onto paper napkins over supper and lay them out on crawfish-chimneyed lawns early in the spring. Young men buy fiberglass.

Down by the shrimp processing plant, tilting into the bayou where Falgout Canal meets Bayou du Large, a developer has gotten hold of a chunk of sugarcane field. Big square fishing camps up on stilts, ready for future hurricanes, dot the cane stubble. Their plastic clapboards are painted suburban America colors and they have green tin roofs. They look strong and foreign.

But the house across the street is Jeff's, not Jim's, and when I get out of my car and crunch across the shells toward it, he stirs up on the balcony, and comes to the rail, leaving the blond woman he has been sitting next to on the old car seat up there that they use for a couch. I feel uneasy under his hardening eyes. I tell him who I am, that I wrote a book some twenty years ago that mentioned his uncle.

Jeff's face transforms like an invisible hand peeling a mask from it. "Why, I remember you," he says, a softer light now in place of the glare. "I was a kid when you was here. I remember that time you came back from Buckskin Bayou. I was standin' right over there and everyone was talkin' about the fella from up north that went dredgin' with James." I don't remember Jeff. I remember a lot of kids scampering around then, mostly Willie Junior, because he loved oysters so much. Now Jeff is twenty-seven with a house of his own—a big one on stilts, a woman, and at least one child, judging from the toys scattered about.

The woman comes to the railing. She is slender and wears a gold-plated necklace and earrings. "I didn't know your uncle was in a book," she exclaims. "Why didn't you tell me?" Jeff looks kind of embarrassed, then irritated, and then tells her that he thinks his aunt Rachel has a copy somewhere. I don't know if she and Jeff are married but she seems to be living in the house. She says she is from Houma, a city of 30,000 about twenty minutes north of Bayou du Large. She says she never knew that Bayou du Large and its little world of oyster luggers, oystermen, and marsh life existed until four years ago when she met Jeff at a dance. Jeff gives her another irritated look for some reason, maybe because he, like too many bayou dwellers, knows that people here don't quite fit into the rest of America. He turns to me and says that he would be out dredging oysters right now " 'cept that my driveshaft on my boat's broke and the fella who said he would fix it is out on a rig in the Gulf."

I remember that happening to Jim one time, and he got a bunch of people from up and down the bayou, and they hauled the boat right out of the water and he made the repair then and there. I wonder if the problem is Jeff, or whether it is something much bigger than Jeff.

Across the road from the crooked houses, on the narrow strip between the shell shoulder, hugged by the bayou with its boats, are the remains of Jim's business, mainly a couple of big fuel tanks with the lettering on them "James Daisy & Sons." A couple of smashed up cars are next to them, as well as odd bits of machines and engines, all rusted and dead. One of the sons is also dead—Dwight. He died of diabetes after James went. Willie is the other son; he broke away from the family and dredges with his own son, Willie Jr. But it's not a labor of love, says Willie Jr.'s wife, Adrienne, later on, as she bounces their serious five-year-old crack child sitting on her lap. It's because Willie Jr. can't do anything else. "Lord, he's tried all sorts of things, but nothing's worked out so now he's back dredgin' with his daddy." The couple's recreation is weight-lifting and every evening after a day of back-breaking work Willie Jr. and Adrienne go to the health club up the bayou with the child, who is not their own but was given to them by a family whose daughter got mixed up with the wrong person. "You should see this child under the sun. She turns black, black, blacker than she is already," says Adrienne. Flattening abs and parenting a child of addiction are new in this otherwise traditional little community poking into the marsh.

I wonder what Jim would have thought. Probably would have liked the child. Probably would have thought pumping iron weird, what with all those oyster sacks out there wanting to be hefted. Jim had a love affair with his oysters, though he liked trapping muskrats, too. "Man, there is nothing better in life," he once told me, "than a dredge full of salty oysters. Lord, pop them open and swallow 'em, an' you think yer in heaven."

Dwight and Willie worked for him, Willie, who was older, in command of a lugger by himself with two deckhands. Dwight worked on his father's boat and wasn't happy about it. They'd dredge the reefs every day out in Buckskin Bayou right in front of their cabin, making lazy circles and crossing each other's bows in a loopy choreography. Every fifteen minutes or so they'd winch in their dredges and dump them full of oysters on the deck. At the end of two days, the luggers would come groaning up Bayou du Large to a big trailer truck that would load them for a shucking house in Florida. Jim made a lot of money from oystering and he worked for it with love. He made $100,000 some years and maybe another $20,000 or so from trapping.

If it were not for the engine, Jeff says, he would be dredging. He doesn't have to go as far as Buckskin Bayou now, he says, and did I know that Hurricane Andrew in 1992 had "picked up Uncle James's cabin and scattered it all over the marsh?" But there wasn't so much marsh left out there because it had all sunk anyway. Now, his oyster reefs are right down Bayou du Large where the houses end. I pictured the place I thought he was talking about, at least the way it was twenty years ago. It had been scrub pasture of palmetto and mangy cattle. The idea of oyster reefs growing where cows had grazed a short time ago seems funny, the kind of funny you feel when something so big is happening, so out of control, that all you can do is shrug, sit back, and watch, or get out of the way.

I guess I crack a smile when Jeff is telling me about his oyster reefs. His face suddenly turns serious. "We got water right in our backyards. Sometimes it come up right under our houses. That's why everybody got stilts now."

By this time a little crowd has gathered, feet scuffing the crushed oyster shells a respectful distance from me, with my head angled up talking to Jeff. Now Jeff descends from the balcony. People up and down the bayou have heard that a stranger—but a known stranger—is talking to Jeff, and they are curious. When Jeff arrives on the ground, they press a little closer. One of them, an old man with a limp and rheumy eyes, blurts out: "Hey, don' you remember me? It's Buddy, James's little brother." He comes right up to me. I am puzzled and I guess I look it. Somehow, I have put him out of my mind. Then I remember him. He was always following Jim around and Jim never paid him much heed except to tolerate him as a family member. During evenings when Jim was home, Buddy used to come into the little kitchen where Rachel puttered, cleaning up after supper, and sit in a corner real quiet. A fixture, no one said anything to him. Rachel just put a cup of coffee in front of him with a spoon. He always stirred it but never drank it. He'd sit there for an hour or so never saying a word but looking around the room endlessly, as if searching for something, maybe just an acknowledgment.

One time he caught me looking at him. He stopped stirring and leaned over the table. A bright light gleamed from his eyes and he said, "You know what I like 'bout them oysters? You look at that meat real close an' you know what it looks likes? It looks like a woman, you know what I mean, a woman down there," and he pointed beneath the table. A crooked grin broke his face like a shooting star and then vanished. Then he picked up the spoon again. No one appeared to have heard him except me, or at least no one took any notice. Ten minutes later, I glanced over at the corner of the kitchen and it was empty.

Now he seems different, more able to stand alone. "James passed on," he says, straight and clear. "I go dredgin' by myself now. I got my own reefs an' I'm doin' real good. Well, you take good care now. You oughta see Rachel. She's up the bayou visiting with Sheena." And he kind of backs away with his limp, making him bob up and down, and then turns and hobbles off to a little house next door to Jeff's.

"That's Uncle Buddy," a little boy says, turning to look up to me. "He's always buyin' me stuff."

Sheena's house is about two miles up the road and as I drive there people by the side of the road wave at me, hesitantly. I know that the telephone lines overhead have been busy. The house is a fat yellow polyvinyl clapboarded rectangle with green shutters. It looks like it should be in a New England suburban subdivision except for the stilts holding it up which allow for a garage at one end and a recreation area at the other and a lot of space for water to collect during floods. A bunch of people are hunkered down at the recreation end over a wide-brimmed bucket heaped with crawfish. I don't recognize any of them but someone offers me a plate of crawfish and someone tells me that Rachel is upstairs with Sheena. A babble of goos, and ahhs and strangled sounds come from a girl strapped into a wheelchair with a neck support so her head won't flop over. I guess that this must be one of Sheena's children, little Caressa. I remember Jim telling me about her on the telephone, one of the last times I talked to him. He told me about "my little girl's floppy baby." "That little girl jus' don' have no muscles, it seems. She's like a little piece of spaghetti, she is." Then he added: "Mus' be the water or somethin'."

Little Caressa is a Louisiana statistic, but not one with any known parameters. As you travel through bayou towns, where still waters are steeping with pollutants dumped long ago by oil companies, you see, more frequently than seems necessary, children with arms and legs and mouths that don't work right. But perhaps these children are blessed in their own way, for in few places other than South Louisiana, where families run large and loyal, are such kids so well and lovingly cared for. As little Caressa oohs and aahs and drivel courses down her splotchy chin and onto her new shirt, various family members pat her on the legs and others wipe her mouth and clean off her shirt. A couple of others coo at her and someone wonders if she needs a diaper change.[1]

Upstairs is cool and soft underfoot from the pink shag carpeting. Sheena and Rachel come out of opposite rooms and both squeal when they see me, though I never would have recognized them. Sheena used to be a shy high school girl who did her best to avoid the man from New York. Now she is the suburban housewife, fiercely proud of her substantial home, of all the cars parked on the grass out front, of her husband, Kenton, known as Coon, who is in the air-conditioning business, of the big TV where just now *Love Connection*, meant to set fantasies soaring, is being wholly ignored. I comment on all the signs of prosperity and a look of pleasure grips her face in a tight, appreciative grin that people put on when they want to let you know that they are possessors of things other people

---

[1] Oil field waste is an endemic problem in Louisiana. In the largely Houma Indian community of Grand Bois, just south of Houma and north of Bayou du Large, one of the largest hazardous waste sites in the state has created years of controversy. Since 1984, a slurry of chemicals including benzene, hydrogen sulfide, lead, and cadmium has been deposited in its eighteen open pits located only three hundred feet from some homes. At present, the facility receives over a million barrels of oil field waste per year. Residents have complained of respiratory and gastrointestinal problems. Medical records show high lead levels in blood samples of children. However, no definitive links have ever been established between these toxins and severe diseases or permanent disabilities.

naturally covet. She asks me if she can fix me a plate of crawfish from downstairs, a soda, perhaps. Sit down, she offers, and points to a huge reclining armchair covered with sunflowers.

Rachel is waiting her turn. She had been Jim's helpmate to a worrisome degree when I had seen her last, her mother's replacement—cooking for him, washing dishes, ironing, grocery shopping—always saying, "Daddy, what you want me t' do now?" and Jim saying something like, "Sweetie, we got to start thinkin' about gettin' some food together for supper. Why don' you see what we got in the fridge, honey?"

Now she has no daddy and she is different. She gives me a kiss and a hug. Her hair is cut in a ball with blond spikes and gold sparkles from her fingers.

I guess she has not married. She stands planted and alone. The first thing she says is that Jim bequeathed her his oystering business, meaning his big lugger and the 800 acres of oyster reefs he leased from the state. She says she hired three crews from "over in Mississippi," who drive to Bayou du Large every morning and are working the reefs as the sun comes up.

"I git to travel a lot now," she announces proudly. "Last year I went to Paris and that's the most beautiful place I ever seen. I went to Morocco one time but I was scared there. I wanted to leave. Thought I was goin' to get lice in that casbah. St. Thomas is my favorite, though; I like the jewelry stores there." She is wearing a big gold ring with about ten diamonds in it and she pulls it off and tells me she bought it in one of the stores there. "Yep, I'm doin' real good. You wan' to see some pictures of Daddy?"

She rummages around in a huge bag draped over her shoulder and extricates a folder of snapshots. There is Jim with his big smile and shining eyes. There he is on the little dock at Buckskin Bayou with a pile of oysters at his feet. There he is beside a big trailer truck with sacks of oysters going up the conveyor belt and filling its maw. It's almost like he is right here is in the room with us, never died at all, just keeps doling out his generosity from wherever he is.

"It ain't the same since Daddy died," she says. Sheena nods. "He was real strict with us and we minded him. Now I don't see

Willie. He's done some bad stuff and I'm not goin' to bother any-more trying to talk to him." Her face changes to sad and I thought she might cry, alert to the realization that her father had held the family together. He'd done more than that, I thought; he'd held a society together.

Outside, a shout goes up from the relatives down under the house as a shiny new red Ford Explorer turns in and eases into place beside the other shiny cars. Sheena rushes to the door announcing that Coon has arrived. He is a little man with quick movements and darting eyes. When Sheena tells him I am from New York, he says he spent six months in College Point, Queens, installing machinery in *The New York Times* printing plant when it was under construction. "Man, I could see them skyscrapers in New York from my apartment, but I never went there. I was too scared. I'd heared stuff 'bout what went on there. I jus' went to work every morning an' went back to my apartment every night. Lord, was I glad to git back here."

Coon and I wander out to the yard in back of the house to set up a trampoline for some of the kids. The stairway leading down is lined with alligator skulls. Coon, seeing me looking at them, says he killed sixty-three of them last season right after he returned from College Point. "I growed up in the marsh an' dredged like Jim an' trapped, an' did some trawlin' fer shrimp, too, but I love hunting gators. I don' have to live off the marsh no more. I got my air business. But, man, if I can earn me a couple of extra thousand dollars a year huntin' alligator, an' all it takes is a week, then I ain't goin' stop doin' that. That's about the only thing that I see the marsh is good for now."

He looks at me a little hesitantly when he says that and I know there is something not right. I haven't met too many people who know the marsh, who have lived in it and off of it, to turn their back on it. I wonder if what he said was for the benefit of a stranger from firmer ground.

Chapter Two

No Marsh, No Oysters

It has become a fact of life that Louisiana is simply disappearing into the Gulf of Mexico—compacting, sinking, drowning, chewed up and swallowed. It is happening to an invaluable and irreplaceable land. One quarter of the nation's wetlands stretch along the coast here, 40 percent of the lower forty-eight states' salt marshes. Of the tens of thousands of acres of wetlands lost each year in this country, 80 percent of the loss is right here. There used to be five million or so acres of marsh and swamp, gradually accumulated over thousands of years in the wake of the Mississippi's lashing back and forth in search of an easy exit to the Gulf. A million have disappeared over the past century. Three million-plus acres remain. But their grip is slipping, a sloughing away that people in Louisiana explain through a variety of astonishing statistics and images. The wetlands are vanishing, some people say, at the rate of 16,000 acres per year. Others place the unraveling of the coast at twenty-five to thirty-five square miles per year, a mass of land comparable in size to Manhattan. Others prefer a more familiar image, a disappearance at the rate of a football field every fifteen minutes. However Louisianans describe the demise of their state's coast, their voices rise in exasperation, their gesticulations and frightened looks impart a sense of foreboding. Like a country pastor, they exhort outsiders to heed the alarm.[1]

People should listen. What is happening in South Louisiana is bound to affect New York, Atlanta, Chicago, and Denver, certainly Washington, D.C. The spontaneous sermons in bayou towns are not vacuous evangelism. To an unwitting extent, this country is dependent on Louisiana. It's not music, Mardi Gras and the French Quarter frolicking at issue; it's food and energy. Louisiana produces 25 to 35 percent of the nation's annual seafood catch, excluding Alaska's. This does not include the 25 percent of the country's annual oyster harvest.

Food is not the only issue, however. Energy is another. The dangers to oil and gas production are scary to contemplate. One-quarter of this country's natural gas crosses the vanishing coastline and deteriorating wetlands through a maze of increasingly vulnerable pipelines. "It's like spaghetti down there," one Chevron engineer tells me, pointing into the Gulf's murky waters. Almost 20 percent of the country's oil comes out of or crosses the state's wetlands.[2]

That's from wells onshore or within Louisiana waters just offshore. Oil and gas production offshore of Louisiana beyond three miles is huge and growing. The states of Alabama, Alaska, California, Louisiana, Mississippi, and Texas account for 25 percent of the nation's offshore production. The three biggest producing states are

[1] There are indications that the rate of loss has somewhat diminished over recent years, from forty to forty-five square miles per year during the 1980s to the present twenty-five to thirty-five square miles. No one knows why; there is some speculation that more regulated canal dredging may be the reason.

[2] Approximately 13 percent of the crude oil that comes into Louisiana is piped from the LOOP (Louisiana Offshore Oil Port), a docking and off-loading facility on stilts twenty miles out in the Gulf. The crude travels through a high-pressure forty-eight-inch diameter pipe to the Cloverly Salt Dome in Lafourche Parish where it is then distributed, by pipeline through inland marshes, to refineries. The pipe coming ashore from the LOOP carries one thousand barrels of crude per minute. If it were to break as it crosses the coast or the marsh, the resulting spill could amount to 2.5 million gallons per hour.

California, Louisiana, and Texas. But production off Louisiana's three-mile limit dwarfs that of California and Texas, accounting for about 80 percent of the entire country's offshore production. All of this energy flows into Louisiana for processing and distribution.

If a slow moving powerful hurricane—a Mitch, say—that killed ten thousand people in Honduras and Nicaragua in 1998 and dumped twenty-five inches of rain in a day—were to cross Louisiana, horrendous damage could result to the nation's oil and gas supply, to say nothing of the people living in and around the wetlands. Wetlands absorb wind-driven high waters; for every 2.7-mile width of wetlands, a storm's surge is diminished by about one foot. Without that sponge, floodwaters have free rein to ravage inland areas.

The farming town of Erath (population 2,500) sits about five feet above sea level in Southwest Louisiana. Surrounded by rice and sugarcane fields, crawfish ponds and sleepy cattle, it is also home to a gas facility oddly called the Henry Hub. Here, giant pipelines arch silverly out of the ground like alien serpents. They spiral on the surface, some joining other lines, others fractionating into a myriad of valves, tanks, and gauges. The Hub, administered by Sabine Pipe Line Company, a Texaco subsidiary, is a focal point for the distribution of natural gas across the country, mostly to the eastern half. Five interstate and twenty-three intrastate pipelines intersect here to send a daily average of 750 million cubic feet of gas from Louisiana to Chicago, Atlanta, Boston, New York, and to points in between. That is a lot of natural gas, about as much as the residents of Los Angeles consume in a year, according to a man named Roy Kass, whose job at the Department of Energy is to survey how much natural gas is being used around the country.

Something else happens at the Henry Hub which is crucial to the natural gas industry, to Wall Street, as well as to millions of natural gas consumers. Buried in the diameter of one of the pipes—a ten-inch trunk pipeline—is a special gauge. It meters all the gas coming into and flowing out of the Henry Hub. The readings set the future price of natural gas, a commodity like wheat and pork bellies, on the New York Mercantile Exchange. Your monthly gas bill is determined in Erath, Louisiana, five feet above sea level and eight miles from the receding coastline.

I asked Kenny Tauzin, one of the Hub's operators—and, he guessed, a fifth cousin of Congressman Billy Tauzin—what would happen if a hurricane blew in. "Why, we'll all just get out of here," he said matter of factly. Then he explained that the Hub could be operated by another facility, "a remote," over in La Rose, a town as low sitting as Erath but a hundred miles to the east. A vision of massive flooding swam into my mind—downed communications lines, snapped pipelines, and frantic brokers screaming on Wall Street. I voiced some of my anxiety to Tauzin but he assured me in a way that made me think that he wanted to assure himself as well that no storm could live up to my vision.

Half a century ago fishermen, trappers, and bayou dwellers began to notice that something was amiss in the marsh. Houses low down behind levees seemed to be standing in water more frequently than the usual floods would warrant. Coastal roads turned into isthmuses bordered by open water. Electricity lines appeared to march out to sea. Such stealthy disaster beguiled and frustrated scientists. They compared recent maps to older ones. What they found verified just what those who had worked and lived in the marsh had been saying: the wetlands were disappearing at an alarming rate.

As in many dilemmas that affect the environment, the causes are far easier to figure than the solutions. Almost everyone agrees that the leveeing of the Mississippi River is the principal cause for the wetlands' sorry state. There's irony here. The Mississippi created the wetlands in the first place; now the river was destroying them, but not without the profound assistance of man. Over the past seven thousand years, the river carved five, some say six, routes to the Gulf, each one resulting in a separate delta. Each spring, floodwaters from the north careened down these courses, flooding banks and sending approximately 200 million tons of sediment into the forming marsh.

The deltas and the sediment created South Louisiana. They remain its foundation, although a deteriorating one.[3]

Any river is an opportunist. It seeks the path of least resistance to its mouth, whether another river or the sea. The delta that it builds proves, over time, to be nettlesome, tons of sediment eventually blocking the route. Current slows as a result, sand bars form, islands rear up, the main current splays into rivulets across the delta. Upstream, the water grows restless, wanting to reach the end of its journey, held back by the growing and impeding delta downstream. Inevitably, the river's current begins chewing at a meander's bank in search of a weak point. During a spring flood, it makes a final escape and cascades down a new course.

A millennium ago, the Mississippi adopted its present course, abandoning its former route down Bayou Lafourche. If humans have their way, the Mississippi will never change its course again, even though it is champing to do so by jumping into the nearby Atchafalaya River. But no one is about to let that happen; engineers have frozen the Mississippi in its channel. Its glorious width and its meanders through the nation's heartland divert attention from the fact that its manmade function is to serve as a huge navigation ditch bisecting the continental United States. Its drainage basin is comprehensive—forty-one states and the central Canadian provinces that flow into the Gulf rather than filter through the wetlands.[4]

[3] While the Mississippi's sediment load is no longer permitted to replenish the surrounding wetlands, it is also true that the Mississippi carries approximately 70 percent less sediment than it did a century ago. The reasons for this reduction are uncertain; the damming of the river's many direct and indirect tributaries is thought to be a major factor.

[4] Some people concerned about the quantity of pollutants included in this load—mainly nitrogen and phosphates and insecticide residue from farms—have drawn the analogy of the Mississippi as a giant sewage system. The analogy can be put on its head, literally, in a quite fascinating and ecologically sound way. If the map of the United States is turned upside down and the Mississippi and its tributaries considered a root system and the river's active delta, a tree's crown, suddenly Louisiana and environs appear in a healthier perspective.

The river's confinement began more then three hundred years ago when Europeans began settling behind the protection of the natural levees that sediment-loaded spring floods had thrown up along its banks. The levees were not high enough to prevent floods from washing over them and flooding the adjacent fertile lands, obliterating settlements and agriculture. The solution was to improve upon nature's work. Building up the Mississippi's natural levees has been an ongoing process as a result, one that continues today under the auspices of the Army Corps of Engineers, charged with the double tasks of preventing the river from flooding and providing a shipping channel for the nation's commerce.

The leveeing, one of the world's most massive and successful engineering projects, has altered the face of South Louisiana and its wetlands forever. Impenetrable walls of dirt and concrete, the levees are causing the land on the other side of them to starve. The river is no longer permitted to deposit its nutrient and silt-laden floodwaters into the wetlands. It is forced to dump its riches far out in the Gulf off the continental shelf. The coastal marsh, bereft of building material in the form of mud, silt, and nutrients, is compacting and sinking. In some places, the land is subsiding as much as 1.5 feet per century.

But that's only half the story. While the land is sinking, the Gulf is rising as the effects of global warming set in. The result is particularly troublesome in Louisiana. Relative sea-level rise—the combination of subsidence and sea-level increase—amounts to almost three feet per century in New Orleans, four feet down the river in Pointe a la Hache, making this part of the Gulf coast a show case precursor of what will happen to the rest of the country's perimeter as the climate continues to heat up. With just a hint of scientific perversity, geologists, coastal experts, and the media eye coastal Louisiana as vultures do when circling a dying cow. New Orleans is regularly called the new Atlantis on TV news shows. Computer simulations show hurricane storm surges coursing through and drowning the French Quarter.

Land loss is also due to the oil and gas industry and the Army Corps of Engineers' ignorance of natural drainage patterns, as canals have been dredged through the marsh in any convenient direction, compromising the slip of water to the Gulf. Barges and tugs traveling these canals send surges of wash into the soft banks, tearing them apart. Canals entering the Gulf invite storm surges, which have not only widened their channels but flooded interior marshes, killing off freshwater vegetation. Maps prepared for a recent study by the United States Geologic Survey of wetlands loss in South Louisiana between 1932 and 1990 indicate that the oil and gas industry has had a huge impact on the coast. In fact, 36 percent of the loss is due directly to exploration, drilling, and transportation while 21 percent is attributed to altered hydrology.

The combined effects of rising Gulf and sinking coast are frightening. Coastal grasslands are falling apart as salt water moves landward, probing up bayous and the Mississippi itself. New Orleans gets its water from the Mississippi, but perhaps not for long. Salt water has crept up the river a hundred miles and is closing in on the city's water intake pipes. Salt has already invaded some rural water wells.

Freshwater marshes, the most productive and diverse for their variety of vegetation, are being particularly hard hit by salt. Grasses with such vaguely disturbing names as sensitive jointvetch, coontail, fanwort, hornwort, smooth beggartick, stinking fleabane, and common frogbit succumb quickly in the presence of salt. If the salt remains, its sulfides block respiration and kills root systems. Plant diversity decreases, eventually permitting only salt grass and cord grass to survive.

The northern movement of salt is inescapably detectable. Redfish, a saltwater species which, along with sea trout, obsesses the state's sports fishermen, used to be rare in Golden Meadow, a town along Bayou Lafourche, twenty miles inland of the Gulf. Now the species is common there. The salt water has crept another twenty miles north clear up to La Rose and boys fishing the bayous near that town regularly bring in redfish.

Corporate America, with a stake in Louisiana, is just beginning to smell danger. Small but telling indications of the growing worry are becoming more common. When a Wal-Mart outlet in Galliano—

another Bayou Lafourche town—closed and put its empty quarters on the market, it turned out to be a hard sell. Potential buyers drove the price down, citing the salt water in wells and the growing hurricane danger as the marsh loses it protective mass.

In February 2000, State Farm Insurance, which carries one-third of the state's homeowner policies, announced a severe curtailment on the writing of new policies in coastal Louisiana because the potential for massive hurricane damage is becoming too great. As Louisiana sinks, as salt water creeps north, America far beyond its reach is just beginning to understand that something is profoundly amiss.

Drastic population spikes and dips of the wetlands' richly diverse fauna are more common now, a sort of fibrillation of an ecosystem. Muskrat populations have plummeted, restricted to dwindling freshwater areas, and even bullied out of these marshes by aggressive nutria. Exploding nutria colonies have turned pockets of the marsh into war zones of mud and floral death. The future of the state's prodigious shrimp production—almost 20 percent of the entire country's—is tenuous, as salt water penetrates northward killing off plant species whose detritus is absorbed by plankton and simple organisms, which in turn are consumed by juvenile shrimp hiding in estuarine fingers splayed over the landscape.[5]

Eugene Turner, a biologist who directs the Coastal Ecology Institute at Louisiana State University, has his students create a visceral display of the importance of the shelter that the marsh provides. He has them tether redfish fingerlings in two places, some in a niche at an estuary's edge and the others in shallow open water a few feet away. Within minutes, the unfortunate fish in the open are reduced to bones and scales, someone's snack. The fish within the marsh's embrace survive much longer.

---

[5] While fluctuations in shrimp populations and harvests are notoriously difficult to track due to the influence of storms, currents, and temperatures, a decade-long study by Roger Zimmerman of the National Marine Fisheries Service suggests a gradual population decline. The reasons may be multiple: wetlands deterioration, pollutants carried by the Mississippi entering both wetlands and the Gulf, or the consequences of climactic change.

Menhaden, an oily fish used in animal feed, is second to shrimp in economic importance to Louisiana, accounting for over half the landings of fish in the state. Its populations appear to be in gradual decline. At least one processing plant has closed. Increasing salinity plays havoc with oyster beds, stymieing growth and allowing schools of black drum and oyster drills to invade. Black drum, a thick-headed fish of substantial size and jaw power that dwells in shallow waters of the Gulf, moves over the reefs when the salinity increases close to shore. Great schools invade the oysters, marauding like garbage pickers at a dump, smashing the shells between their jaws and gulping down the meat. Oyster drills also slither shoreward. Lovely periwinkles in appearance, they position themselves over an oyster with the sturdiness of a Jack boat to drill through its shell and suck out the contents.

And this once pristine environment endures the prospect of experiencing another death, one that threatens not only millions of lives but a substantial part of this country's infrastructure. Like a recurring nightmare, storms and hurricanes barrel across the Gulf and slam into the marsh. Once, soft grasses and pliable earth calmed storm surges and absorbed floodwaters. Now, the diminishing landscape is more accommodating to the furies. With the marsh in fraying pieces and open water broadening, with barrier islands in slivers, wind and waves find fewer obstacles in this flat land. Altered nature has given both freer rein to ravage the coast, where 70 percent of the state's 4 million people live. Ninety percent of them dwell less than three feet above sea level. People talk nervously about "The Big One" coming, as if the next hurricane to sweep the coast will assume the dimensions of a spiritual force not to be denied.

With little to blunt "The Big One's" passage, the dangers include more than the loss of human life. Storms now have leeway to topple oil rigs, destroy the 20,000-mile-plus maze of pipelines that zigzag across the coast carrying gas and crude oil to refineries, to wash out roads and railroads, and to fill the basin that the city of New Orleans occupies, drowning its citizens, its history, and eclectic cultural melange in a horrific inundation of mud-fortified water.

The summer of 2000 brought another ill, as though the Gulf coast is facing a curse. Suddenly, pockets of marsh from the Mississippi River to Texas began to turn brown. As the hot weather progressed, the pockets turned to great expanses. No one had ever witnessed such magnitude of death. The smooth cord grass and oyster grass just gave up, it appeared, turned brittle and crumbled to dust in your hands. There was no sign of disease, even under the microscope. No one knew the cause but a lot of people bet that the two-year drought that had plagued the Midwest and the South had a lot to do with it. The same kind of brown began appearing in salt marshes in South Carolina and Georgia, both drought-stricken states. The brown went deep below the surface. It killed the roots; the grass was gone. Dead roots fell apart like dry spaghetti under the lashings of a hurricane or tropical storm and thousands of acres of mud were swept into the Gulf.

The future is not bright for this coast, and it is peculiar that while so much political and economic attention has turned to saving the Everglades, the Chesapeake, the Great Lakes, and old growth forests, few people outside of Louisiana understand the destruction and its implications going on here. Few people, even in New Orleans, understand it, a city more enamored with experiencing good times than attending to brewing crises. Too, a way of life is disappearing, perhaps profoundly the saddest part of this story, for it affects those who have lived close to the land for generations. People who have built their homes up and down the bayous, who have tied their luggers and shrimpers—their means of supporting their families—almost literally to their doorsteps, are finding that the marsh no longer sustains them with the same predictability that their forebears enjoyed.

Louisianans have long taken advantage of the state's mixed economy—working for stints in the oil industry as roustabouts and welders between pursuing the traditional activities of fur trapping and shrimping. That they must turn away from the marsh—the footing of their lives and traditions—contributes to a malaise coursing through the bayous. The petrochemical industry, devoid of family solidarity, empty of stories to be passed down through the generations, lacking in heroics, focusing on work for pay, is increasingly their permanent god, soulless and mechanical. There is potent irony in the fact that while the Cajun food and music emanating from South Louisiana now dot the American landscape, their very origins are vanishing in a morass of mud and decay.

While the leveeing of the Mississippi, its tributaries, and its distributaries to narrow beds is responsible for much of the damage, some insults to the ecosystem always receive extra vituperation. The Mississippi River Gulf Outlet is everyone's whipping boy. It's a seventy-six-mile canal dredged in the late 1950s by the Army Corps of Engineers to enable container ships to travel a straight line from the Gulf to New Orleans rather than having to spend ten hours winding up the Mississippi's tortuous route. The Corps dredged it 500 feet wide. Now, erosion from ships and storms has gouged it 2,000 feet wide and made it a freeway to New Orleans for any hurricane that happens to come from the right direction.

In all fairness, the Corps, in its recent, much-flaunted reincarnation as environmentally friendly, realizes that dredging the MR.GO., as it is awkwardly and often referred to, was a big mistake. The Corps has made recent moves to close it to shipping, though the motives are driven by economics on two fronts, rather than concern for the environment. First is that container ships, which the MR.GO. was created for, have grown so large that their draft exceeds the thirty-six-foot depth at which the channel is maintained. Newer container ships are thus forced to wend their way up the Mississippi, making the MR.GO. obsolete. And second, the positioning of the canal facing the Gulf makes it prone to rapid siltation. After Hurricane Georges blew through in 1998, the Corps had to spend $35 million clearing away the silt, which calculates at $48,000 for each of the 730 ships that used it that year.

Like a thumb drawn through soft butter, the MR.GO. cuts across a valuable freshwater marsh and four natural levees, remnants of a former route the Mississippi followed to the Gulf. The surrounding marsh, now vulnerable to storms and salt water, has all but died as a result, along with 40,000 acres of mature cypress trees. Now, storm surges can invade the marsh through the straight-arrow channel and smash into New Orleans. Intruding saltwater has killed off the marsh's muskrat, crawfish, and diverse grass species. And the flooding that the channel causes has dislocated thousands of people.

Many of them are, or used to be, Junior Rodriguez's constituents in St. Bernard Parish. Embittered after twenty-six years of doing battle against the Port of New Orleans and the Corps as a Parish councilman, he says that "closing the MR.GO. is the biggest challenge in my lifetime." An immense man with a barrel chest that produces oral thunder, his political career is lodged in distrust of the Corps and the Port of New Orleans's administrators. He tends to explode with colorful language at meetings about the MR.GO. At one I attended at Corps headquarters in New Orleans, he faced the engineering autocrats like a mountain. "The only reason I am sitting here is to see the MR.GO. closed. This thing has destroyed the parish. We got no swamp left; we got no marsh left. It ain't an option for the people of the parish to keep this open."

People are used to Junior's bolts. Someone tells him to hush up a bit. He turns like he might squash them, but then gives them a little smile and a nod.

Junior took me out into his parish's sinking marsh one day in a boat driven by his friend, Gatien Livaudais, a landowner who has seen more of his land disappear than he cares to think about. We were going to meet Gatien down at Penny's Café along Judge Perez Drive. We left from the St. Bernard Parish Government Complex where Junior's plaque- and photograph-studded office is as big as many peoples' homes. On the way in his tiny Mazda into which he angled himself in sections, he told me that he is an "Isleno." His ancestors had come here from the Canary Islands in the early nineteenth century with "a mule, two shovels, and a land grant," and settled on what was then pristine marsh. As we passed mall and subdivision after mall and subdivision, he commented on the distance the people of St. Bernard—many of whom call themselves Islenos like him—have put between their lives and the marsh. "Apathy, that's what I see. 'He' (the generic citizen) only cares if his trash is picked up. Otherwise, he's content. 'He's' not worried about land loss. The only time 'he' worries is when a hurricane comes." People are worrying more now. "It used to take a while for a storm surge to get here from the Gulf," he adds. "Now, with the marsh gone, it's here in hours."

At Penny's, a jumble of restaurant and general store surrounded by a crushed oyster shell parking lot, the woman behind the counter looked up as Junior came in. Some men sitting on stools with their back to the door asked her who had entered. "It ain't a customer, I'll tell you that," she said jovially. "It's shit. That's what come in here." Junior accepts this kind of talk as much as he dishes it out. One of the men was Gatien. With his wire-rimmed glasses, carefully clipped gray hair, and L.L. Bean-type outfit, he looked more like a New England college professor than the owner and overseer of twelve thousand acres of marsh. He set me straight about his origins in about a second—at least I guess that is what he was doing—by informing me that he shared his first name with a cathedral in Tours, France. Forget New England. His family, with thick French roots, has owned marshland since 1904, back when muskrat pelts brought a good price. That is how his people supported themselves. Now, there are no muskrats, driven out by the encroaching salt water.

Minutes later, we were cruising down Violet Canal toward the MR.GO., the marsh on either side of us pockmarked with erosion and sinking. I asked him to point out his land. He turned to me with a bemused look and said: "Why, as far as you can see, it's all mine." I could see to the horizon to the north and south. But then he added that he had lost thousands of acres to erosion. "There's no recourse when that happens," he said.

Junior, up in the bow, was being uncharacteristically quiet. He was hunkered down, wearing a solemn face. The marsh seemed to calm him, or make him still with anger, I couldn't tell. Even when we arrived at the MR.GO., he did not say much. The canal is actually not very interesting, unless perhaps you have watched it grow from year to year, chewing away at the marsh and now so wide that it runs through one edge of Lake Borgne. To me, it looked like an unnatural act, a trajectory driven through the wetlands as straight as a bore through a coal seam. Not until we arrived at a particular place down a bayou that angles off the canal did Junior perk up from his profound slouch. Towering trunks of thousands of dead cypress stuck up from the marsh like giant toothpicks. It is a lonely place, eerie in disease. Even the silhouettes of a gaggle of cormorants on the dead branches of a tree instilled no spirit of life here. "It's like an environmental graveyard out here," huffed Junior. "It's a monument to the marsh's death." Then he went quiet.

Efforts to make things right again in South Louisiana are unsubtle. It is one thing to have the environmentalists scream about abuse, it is another to have a pro-business politician such as Billy Tauzin try to draw attention to wetlands loss. A longtime, down-home Republican congressman (formerly Democratic—a switch that has resulted in some career angst), Tauzin actually called a conference on wetlands preservation on October 2, 1998, shortly after Hurricane Georges had blessedly spared New Orleans by forty miles, smacking into Mississippi and Alabama, and a few weeks before Hurricane Mitch would slam into Central America. Georges had been on track to make a direct hit on the city, sidestepping to the east at the last minute, not the first time that had happened, almost as if hurricanes take pleasure in fooling with the nerves of New Orleanians.

The near-miss scared the bejesus out of a lot of people and a lot of them became oracles overnight, predicting that Georges was a warning, that the next one was going to strike head on. Playing soothsayer is tempting now, with the planet warming and hurricanes on the rise.

The fear was Tauzin's cue. For an entire day, wetlands experts, state politicians, even Governor "Mike" Foster himself, and Army Corps of Engineers dignitaries from Washington, D.C., held forth in the auditorium at Nicholls State University in Thibodeaux, ruing the loss of Louisiana's marshes. Self-blame became their stigmata. "We have met the enemy and it is us," their chant, and cries for financial help, their group wail. Recriminations mounted. "Clean up your toy box," someone yelled out rhetorically to the oil industry for messing up God's plumbing of the marshes with canals going every which way. Someone else wondered how the rest of the country could take Louisiana's plight seriously, observing, "We have a national reputation for dumping chemicals in the marsh." Another observation quickly followed about the fact that Louisiana has more oil spills than any other state. So perhaps it's no wonder, someone concluded, that the federal government pours $100 million each year into restoring New Jersey's beaches for sun worshippers and property owners, but sees fit to fork over only $40 million per year to save one of the country's greatest wetlands.

Observers of and players in the environmental scene in Louisiana always agree on one thing: wetlands loss is someone else's or something's fault. And if you can blame the Corps or the oil and gas industry or the levees or the canals or just rising sea levels hard enough, you can make yourself and others feel better for a while, even though nothing gets fixed. Today, blame was being passed around like hot coals. But there was a difference in reaction, maybe helped along by Hurricane Georges. People in the audience seemed to sicken quickly of hearing the same old blame, and the denials that always followed. Tension was rising. The time had come to put a stop to all the talk.

It was the engineering industry that brought a halt to the talk. A surprise, for its members were the very people who had dredged all the canals willy-nilly in the first place, before anyone cared. Steve Smith rose to his feet to speak and the audience hushed up. An engineer himself and an executive with T. Baker Smith & Sons, his family's engineering firm, he broke through all the rhetoric. A young man, he's a good old boy from Houma who hems and haws and shuffles. A lot of people at the conference know him and he is obviously well liked.

There was silence. Steve Smith hulked up on the stage behind a lectern. Then he sucked in a volume of breath and exploded: "It doesn't matter if the land is sinking or the water is rising, we still have to get taller boots. Com' on, let's do something."

The audience broke out in cheers and screams and thumping applause. Steve Smith had brought them back to reality. All the talk and finger pointing in the world wouldn't make the problem go away.

Hope had to return; there was no other option but to bring it back, a kind of optimism by necessity that is endemic to South Louisiana where every day, no matter how much manipulation goes on, nature is still going to be in your face. The engineering firms helped; there were half a dozen of them outside the auditorium. Their lavish poster displays, videos, and brochures boasted their expertise in resurrecting the dying marsh, planting new marsh, removing spoil banks that have blocked nature's hydrology for half a century. Here was a can-do group who could save the coast—for a fee, of course—and never mind that their earth-moving in years past was partially responsible for wetlands destruction.

Congressman Tauzin helped out, too. Whatever people may feel about Billy Tauzin's environmental record, he's a consummate politician, and in Louisiana politicians who connect with the people are respected, and reelected. He's serving his eleventh term now from the Third Congressional District, which includes a lot of marsh. Some years ago, the American Farm Bureau awarded him its annual Golden Plow Award, an honor that carries implications, like the clout to turn wetlands into farmlands. Now, constituents are beginning to understand that without wetlands, their lives will be in shambles. Tauzin is right there with them.

Fifty schoolchildren clinched the connection between marsh and continuity of the good life and, of course, Tauzin's continued success. They were marched on stage where some of them were asked to read letters they had written to President Clinton proclaiming their love for the marsh. Ashley Williams, a seventh grader, stammered out, face buried in a piece of paper: "It seems every time I go fishing, more of our land has disappeared." Eighth grader Andy Depret didn't mince words: "Mr. President," he read from his letter, "If we don't stop this erosion process now, you know where we will be. How would you like it if you lost an area the size of Rhode Island in your backyard every year?"

The effect was visceral. You could feel the audience go soft as the reading continued. There were a lot of mushy people in the auditorium by the time the kids marched offstage.

Then an oilman stood up. He was R. Michael Lyons, manager of regulatory affairs for the Mid-Continent Oil & Gas Association, and he reminded everyone, "We have been talking to ourselves for about twenty years" to little avail, as the wetlands crumble. Now, the challenge is to figure out how to get the nation interested.

No one said much after that. Perhaps the years of denial were finally coming to an end. Perhaps Louisiana's isolation in the face of its grand disappearance—a national dilemma—was finally going to result in action instead of complaints.

At first glance nothing about south Louisiana seems to be environmentally amiss. That's a big problem. The marsh exclaims natural exuberance and fecundity. Stand on a levee of the Mississippi down around Myrtle Grove and look west—the waving grass rolls toward Texas as though there's nothing else in the world, dotted here by stands of live oak, there by meandering bayous. Off in the distance, shrimpers and oyster luggers appear to float over the grass as they glide along bayous toward the Gulf. Travel down any waterway here and the bird life is extravagant. Herons line banks in watchful poses, stabbing foolish frogs and minnows with unnerving speed and detachment. In the fall, marsh ponds explode with waterfowl that rise in chaotic clouds before forming great spirals that darken the sky like sheets of rain. Twenty percent of North America's waterfowl pass through Louisiana's wetlands. Like much of the American landscape, from the rolling forests of New England to the preserved interstices between housing developments in southern California, everything looks so damn well put together and deceptively healthy.

Chances are that casual tourists or conventioneers won't leave Louisiana thinking that they have just passed through an environment in shambles. Chances are that if they sign up for a half-day bayou ramble with a New Orleans tour operator, they will be stunned by the area's natural heritage. They file onto the pontoon boats, brimming with pleasant foreboding at the prospect of entering the wilds. They are informed before boarding that the water moccasins, alligators, and snapping turtles won't hurt them "unless, of course, y'all fall over da side, or I wreck da but or somet'ing like dat," as the Cajun-accented operator of a swamp excursion run by Cypress Tours put it to a cluster of State Farm Insurance employees who had been given the afternoon away from their convention in New Orleans. Along for the ride, I am struck initially by these Midwesterners'—from Oklahoma, Kansas, and South Dakota—school-kid expectancy. From the tourist-ready cuteness of the French Quarter to a rickety wharf on the edge of a bayou lined with tilting dwellings, working boats, and working people whose lives are so honed to the environment that they can "hear dem crawfish movin' in da swamp 'bout a mile away" is a leap into a different dimension. The State Farm people behave like disciplined children as they enter this strange world.

Captain James Camardelle says he can't "guar'tee no 'gators" on this chilly April day, but says he has brought a bag of marshmallows just in case. As the boat slides down the bayou deeper into the swamp—past moss-dangling cypress trees and their strange knees, past patchwork fishermen's huts where life must be so alluringly simple before you really think about it, past shrimpers coming in with their nets gossamer like spirits captured in the breeze, past great blue herons and egrets posing better than garden sculptures—that's enough for this crowd. But when Captain Camardelle states with quiet certitude, "Dey got some turtles on a log over der," the whole boat tilts to one side in the excitement of a sight that, truly, is not that grand. But it is grand here, in the midst of this Jurassic Park, which is not fake at all but does seem out of place in this country of neon malls and instant environments.

When Captain James, as he came to be anointed, points out the first 'gator, the passengers don't move. They stare and stare as though they have seen the truth. The 'gator stares right back, a glistening-eyed ornament that comes to life only when marshmallows arrive a few feet in front of its nose. The boat tilts again as everyone gets a good look. Then Captain James offers the strange tidbit of information that the passengers enjoy as much as the 'gator enjoys the marshmallows. It is that alligators that have eaten a lot of marshmallows taste like marshmallows if their tails happen to end up barbecued and served on the end of a skewer.

You have to get up with the eagles that soar over the marsh to see that things are not totally fine. There are nine of us in the Sikorsky S-76 taking off from a patch of lawn outside of the Army Corps of Engineers office complex in New Orleans. We rise and turn as if suspended on a string and then whup-whup-whup along the brown waters of the Mississippi traversing the crescent that supports the city's sinking foundations. In no more than five minutes, the marsh is beneath us. When the chopper banks, I am hanging over the primordial ooze. If it were not for my seat belt and the flimsy door, I would tumble out, a thought that does not scare me in the slightest. So soft-appearing is the marsh beneath, so textured in its early spring green, it looks like a wonderful old blanket on a bed hastily made. In fact, I am under the delusion that if the chopper were to fall out of the sky, no one would get hurt. We would all just plummet down and then bounce back up again. It would be lovely.

What is below, in fact, could be God's bed, well slept in and no effort made to neaten it up. And huge; to the west its sheets are rumpled right to the horizon, its frayed, greenish blanket tattered and hole-ridden in a million places. The vastness and us buzzing around above it make me think of a fly zigzagging stupidly in the rising heat of morning sun.

I have tagged along on this trip to see the vanishing marsh from the air—to affirm for myself that it really is vanishing—an opportunity that I have never had despite long wanderings in the marsh by boat or back roads. The passengers are young congressional aides who flew in the previous night from Washington, D.C. They are being shown a handful of efforts to restore wetlands and barrier islands that various state and federal agencies have cobbled together in an effort to save South Louisiana. There are four of them, young kids pretty fresh out of college, each working for a congressman or senator who has an interest in coastal restoration. Their job today is to take notes and report back to their bosses, who will then decide if they want to vote to send more money to Louisiana for wetlands "remediation," in techno-lingo.

They are awfully quiet. Only one of them has been to Louisiana before. One comes from San Francisco, one from the Chesapeake Bay area, one from D.C., and the fourth, from Vermont, of all places. They must be thinking that they are in a different country. In many ways they are.

The jobs of the other people on board, besides those of the pilot and copilot, are to show these young people works in progress, rescue efforts, as proof that the marsh can and should be fixed. There's Jack Caldwell, the cherub-faced secretary of the Louisiana Department of Natural Resources, Rick Hartman, a National Marine Fisheries Service biologist, David Fruge, a supervisor from the U.S. Fish and Wildlife Services, and Sue Hawes, an environmental project manager of the Army Corps.

I cleverly arranged my entree into the chopper to sit between Sue and a window. With curly gray hair and wire-framed glasses, backpack, binoculars, and jeans, she looks more environmental than Corps. She told me earlier that she and her husband used to teach biology at the college level. Her husband, a bicycling enthusiast, gave up academia at some point and opened a bicycle shop. This sketch appeals to my appreciation of quirkiness and independence in contrast to the stereotypical engineer mentality for which the Corps, of course, is known. I think Sue may be wise, that I might be able to learn something special from her.

We are heading toward one of the eighty-plus "CWPPRA projects," as they are known, scattered across South Louisiana in an effort to keep the marsh alive. CWPPRA—pronounced *quip-pra*—stands for Coastal Wetlands Planning, Protection, and Restoration

Act. CWPPRA is also known as the Breaux Act after its principal author, Senator John Breaux of Louisiana. Passed in 1990, the Act's purpose is to provide federal funding to coastal states for the restoration of vanishing wetlands. Louisiana gets the lion's share—around $40 million annually. Through a complex and quasi-concealed political system, restoration projects are conceived each year by representatives of five federal and state agencies and the governor's office and, supposedly, the "community at large"—landowners, local governments, scientists, and anyone else. Known in Louisiana wetlands preservation circles as "the task force," these roles are today being fulfilled by Caldwell, Hawes, Fruge, and Hartman.

The Naomi Siphon is the project we are angling toward, a marsh restorative that these people are proud of. It's a blueprint for survival, a tiny effort to resuscitate the dying marsh, suffocating under a burden of saltwater intrusion. On the river's western shore, a parallel series of giant pipes hump over the levee. Two-thousand four-hundred cubic feet per second of mucky brown water from the river courses through them into the marsh or what's left of it. What I see is that not much is left of it—just tattered sections of the blanket—clumps, tufts, gnarls, and water-logged morasses. But there's hope. Where the river water's dirty fingers extend through this skeletal matrix, tenuous life and its tissues in the form of grass shoots—mostly bull tongue along with wax myrtle—are beginning to cling to the edges of the new phalanges.

It all seems so simple from a height of two hundred feet. Below is raw creation, man copying God's—or whoever's or whatever's— work, over the past seven thousand years. Below us, nature used to just let loose floods every spring that covered the marsh with silt and nutrients, enabling its grasses to weave a tight blanket to nourish and protect all sorts of life. As products of nature, humans become self-conscious in their efforts to imitate nature's spontaneity. They label things as if to excuse them, hence the unwieldy full name for this little imitation of nature—Naomi Siphon and Outfall Management—which at a cost of almost $2 million is supposed to create some 1,500 acres of healthy freshwater marsh at the end of twenty years of operation.

People imitating nature are sometimes the only logical choice to get things back on track. But that happens only after people come to grips with the consequences of ignoring nature. It was just upstream from where we were wheeling around like flies that people in New Orleans in 1927 decided to imitate nature before disaster got the upper hand. Floodwaters rampaged down the Mississippi, resulting in the worst flooding upstream ever known, so vividly described by John Barry in his book, *Rising Tide*. No one wanted to lose New Orleans. In a sharp bend of the river fifteen miles south of the city, where the runoff butted against the levee, and the floodwaters were backing up and threatening to inundate New Orleans, the Mississippi River Commission (predecessor of the Army Corps of Engineers on the Mississippi) set down dynamite and blasted the earthworks to smithereens. Millions of tons of water surged into a marsh that even then was beginning to show some early signs of starvation.

Best thing that ever happened to the marsh, although the people flooded out would fervently debate that issue. When the floodwaters receded, the marsh had a new coat, layers of mud that soon sprouted healthy green. Hurricanes have the same effect; they resuscitate a starving marsh with the muck they pour into it. Most important, when hurricanes or floods are on the loose, people understand that their power to control nature is almost, but not quite, as powerful as nature itself.

The lesson learned from the big hole in the levee is ongoing just about at the same site that the Mississippi's levee was dynamited. It's the Caernarvon Freshwater Diversion Structure—another awkward name that pretends to give humans some mastery over nature's designs. A bigger version of the Naomi Siphon, it is designed to release 8,000 cubic feet per second of Mississippi water into the marsh. The Army Corps, which built it at a cost of $26 million between 1988 and 1991, hopes to preserve 16,000 acres of marsh by forcing freshwater to push encroaching saltwater back into the Gulf. Sport fishermen love it because the drop in salinity has brought back speckled trout and largemouth bass. But oystermen hate it because all the freshwater flooding over their reefs has killed their oysters. The mollusks have suffocated in the mud or for lack of sufficiently saline water. Oystermen got so mad—and they along with shrimpers are united in Louisiana like no other fishing industries—that they sued the state, seeking a reduction of the flow of freshwater into the marsh to a fraction of what it could be. More pertinently, the courts ordered that the flow be reduced enough so the water would not carry sediment, the marsh's foundation. In late December 2000 a Plaquemine Parish jury awarded almost $50 million to a handful of oyster farmers who claimed that sediment and freshwater had ruined over 3,000 acres of their oyster beds. The decision gave the area one of the highest values per acre of oyster lease ever recorded, over $21,000 an acre. A final ruling is pending on a claim lodged by an additional 125 oyster farmers for the ruination of another 30,000 acres. In the meantime, a lot of marsh that could have been built has not been.[6]

Some of this would-be marsh is, in fact, entirely open water, an enormous crooked square stamped out of the existing marsh as if by a supernatural cookie cutter. It's called Big Mar, so named, some say, as a joke, as in "big sea," in the middle of the "marsh." Actually, this acreage of open water was scoured out of the marsh by the 1927 levee dynamiting, when floodwaters surged through the break. Now, water from the Mississippi flows directly into Big Mar, taking more or less the same route through the Caernarvon Freshwater Diversion.

One late winter day about six months after buzzing over it by helicopter, I was riding through Big Mar with Lonny Serpas, and his assistant, Albertine Kimbal. Albertine refers to herself either as "Swamp Queen" or "Duck Queen," depending on whether she is in a cypress swamp in her airboat or hunting ducks, both volubly expressed passions of hers. Lonny and Albertine work for

---

[6] The state Department of Natural Resources hopes to avoid future legal wranglings with oystermen as a result of water diversions by compensating them for lost oyster reefs before diversions are installed or by offering acreage in comparable areas, which will then be seeded.

Plaquemine Parish which is where the diversion is located. Their job is to supervise its operation. This means that each morning, Lonny has to mount the massive gated concrete bulkhead that separates the river from the parish and raise or lower the gates, depending upon the river's level. His pride in its operation is obvious. He shows off its gauges, valves, and buttons like a boy with a new toy fire engine.

But for the past six months, he hadn't had much to do. The river had dwindled to such low levels that the water didn't even approach the lip of the gates, which meant that no freshwater was getting into the marsh. Nature had put the Corps's marsh rebuilding plans on hold.

Nevertheless, evidence of marsh building abounds. Lonny seems more than happy to take me out into Big Mar in the Plaquemine Parish boat. Albertine comes along mostly because she likes to be out in the marsh showing it off to a stranger. And she likes to watch the movements of waterfowl. What you see out there in the open water is the beginning of creation, at least marsh creation. The levees that border Big Mar, once narrow and abruptly falling off into the water, have widened out to form flats onto which vegetation clings, the beginning of the 16,000 acres the Corps is planning on restoring.

This may be wishful thinking. The Corps claims that exactly 406 acres have been created in ten years of operation in a sampled area that once contained almost 2,300 acres of wetlands. The diversion's life span is supposed to be fifty years for a gain in the sampled area of forty acres per year and, over fifty years, of two thousand acres, suggesting an eventual shortfall. Still, Lonny and Albertine are happy, exclaiming over the mud flats. "It's real dramatic to see what is going on out here," Lonny yells over the roar of the boat's engine. "We need this rebuilding to save us from sinking right into the Gulf."

The one thing that is very wrong on the helicopter is that someone forgot about the overwhelming noise. The congressional aides look down on the Caernarvon Freshwater Diversion Structure below us with mild interest; I have no idea if they are aware of the significance of the restorative effort. Passengers on most see-and-tell trips are given a blow-by-blow explanation of what they are seeing, piped to them through earphones. Attached microphones allow questions and answers. But not here. Someone forgot to bring along the technology. We all discover the strangeness of being deaf and mute. I quite enjoy it, actually, and am able to focus all the more on the disturbed land and waterscape below. Sue points things out to me but I don't know what she is pointing to. We begin a handwritten message correspondence.

"Why isn't more freshwater being pumped into the marsh?" I scrawl in my notebook.

She scribbles back: "It's oyster city down there but too much freshwater kills them." Then she continues ominously in shaky script. "But when all the marsh is gone, there'll be no more oysters."

"Why are there so many dead cypress below us?" I write as we pass over a forest of huge cypress skeletons, not realizing then that it is the same forest that Junior Rodriguez was mourning when he took me out with his friend to survey the vanishing marshes of St. Bernard Parish.

"Because saltwater has encroached here as the marsh has sunk and cypress can't tolerate saltwater."

Then she points toward the horizon to a part of the marsh that looks like some of the tattered rugs I have in my house which should have been thrown out long ago. Below is mostly open water, striated by lines of spoil banks, the residue of oil and gas company dredges that tossed mounds of muck to the sides of canals dug out for pipelines, or to transport drilling equipment into the marsh for wells. The spoil banks sprout lines of runty willow trees that outline the courses of the canals. You can tell which canals were dredged for oil wells. They end abruptly, the banks tracing a shape that loosely resembles a keyhole, hence the name for them—"keyhole canal." I can see probably twenty of them below us but there are no active wells, long exhausted.

With the passage of years, storm surges, countless boat wakes, and just plain time have chewed away at the canals, widening their banks, permitting more and more water to enter from the Gulf. At the same time, many of the canals are at angles contrary to the slight pitch of the land. Their spoil banks disrupt the southward flow of water, trapping it in shallow reservoirs, drowning the marsh beneath it. Some marsh grass roots penetrate a yard or so in the mud. It took a century of river silt to build up that depth. When water covers roots, plants die within weeks. Roots, which hold the muck in place, decay. Wave surges from the next winter storm or hurricane wash away the unanchored muck like a hose let loose on a newly seeded lawn. It's horribly simple.

Sue scribbles: "The Corps would *never* allow such dredging now." This is largely but not entirely true. An estimated ten thousand miles of canals have been gouged out of the marsh over the years. Every year, dredges still gobble up about one thousand more miles, though with far more consideration for the environment.

We've been in the air for three hours, halfway through the three-hundred-mile trip, to view fifteen or so CWPPRA projects, to show the aides how much progress has been made in saving the marsh and how much work there is to be done. We are at lunch at the Kajun Sportsman restaurant in Port Fourchon—fried oysters, shrimps galore, soft shell crabs, po' boys, cole slaw and the ubiquitous iced tea. Jack Caldwell sits at one end of a long table and expounds on different efforts to save the marsh that are funded by the state. Sue is watching him like a hawk, probably to make sure the Department of Natural Resources does not take credit for anything that the Corps is responsible for. Caldwell eyes her, too. Sometimes, she corrects him when he mentions the wrong bird species or gets his islands or marsh grasses mixed up. They seem to have a long-running rivalry.

Now up in the air again, bellies full, we are feeling cozier with each other, which has nothing to do with conversational intimacy, even though we are now wired to newly found headsets so we can hear each other and talk. The noise is still debilitating, penetrating, and exhausting, so alien in such proximity to the beauty below. We are through being startled by the destruction beneath us. No longer voyeurs, we have become critics, students, and instant philosophers. We look down at the ruthless excavations but also at the seemingly endless wetlands that spread to the horizons. Of course, we cannot see the million-or-so acres that have eroded away or sunk over the years. Nevertheless, the disease that infects so many Louisianans strikes us, too: it is the vastness below, the extravagance of life that make the scars appear like cat scratches across smooth skin. Not all people here worry about the coast's demise; there's so much of it left.

Here, far to the west of the Mississippi's dirty outflow, the Gulf's waters are greener and less laden with sediment. The surface seems bouncier, the north wind ruffling tops of the swells with white filigree. We are heading out to the crucial and fragile barrier islands—East Timbalier, Isle Dernier, Raccoon, Point au Fer—to view efforts to save them, and then over to the Atchafalaya Delta where dredges have created thousands of acres of new marsh. Below us is Raccoon Island, a gem to be shown off. Ringed by beaches with an interior of scrub on which great white pelicans loaf and beautiful roseate spoonbills glide over, fourteen months of engineering saved it from a path to extinction, after Hurricane Andrew blew away almost half the one-hundred-acre-plus island in 1992. Two breakwaters were erected on the east and west ends, and now sand is accumulating around both of them and building the island anew. A success story for Congress.

It occurs to me that there are people in my presence who are playing God—siphoning water here, moving tons of muck and sand there to create virtual marsh, hewing up breakwaters to redirect currents. The handout we have been given lists the following for East Timbalier Island:

> Over the last one hundred years this island has been breaking up and moving north. Attempts were made to protect it by placing large rocks on its shore in the early 1970s. The island has continued to erode away from the rocks. [One project] is to create 134 acres of marsh from material dredged from the bay. At the end of twenty years, 37 acres should remain.[Another project] includes restoration of this island by building 131 acres of marsh. At the end of twenty years, sixty-one additional acres should be present. The fully funded cost is $5.7 million.

It would seem, from our tour, that the restoration of South Louisiana is under way to good effect and at considerable expense. Yes, a helicopter flyover seeing both sores and healing tissue, highly selected, would make that impression. Chuck Melton, the pilot, sets the chopper down late in the afternoon at the New Orleans International Airport so the aides can make their flight back to D.C. They look dazed as we all are—by noise, craning necks downward, but also by the constancy of the view below, particularly during the uninterrupted leg from the Atchafalaya Delta to the airport. On the way Melton took a turn around the construction site of Davis Pond Freshwater Diversion, a massive Corps effort to divert Mississippi River water into a part of the marsh that is in particularly bad shape. The project design predicts that some 30,000 acres of marsh will be saved or created over the next fifty years. There had been an hour of flying time with nothing but raw nature beneath us, and only sparse evidence of human efforts to conquer that nature. There are flooded and abandoned farms down there bordered by levees, square fields so subsided that they will be underwater forever. Tractors encased in rust looked like some horrendous experiment gone awry.

The sights sure have been impressive and certainly help Louisiana get a chunk of money. But obvious observations and questions loom. The land is sinking and the Gulf is rising and hurricanes are on the horizon. How can Washington's meager handouts possibly stem the destruction of this delicate land? Why isn't anyone doing anything substantial? These projects we have witnessed from the air appear as mere Band-Aids of mud, rock, and hubris in the vastness of this wetlands in dissemblance, short shrift for such a cradle of food and energy and natural beauty.

Chapter Three

Dying and Resurrection in South Louisiana

Shea Penland is planted on the beach off Marchand Bay, the spring sun beginning to come down hot and merciless, about to turn his winter-white legs, now in shorts, red, and the marsh behind him, green. His brow is already glowing pink, furrowed, and angled toward the dirty sand at his feet. Raked clean of trash, the tilting expanse is tine-grooved with runnels paralleling the gentle waves breaking on the shore. Cathy, his wife, stands a few feet away holding a wad of maps and satellite photographs. With dark glasses, tanned skin, bright smile, she appears excitable, prancing her feet about in the sand in contrast to her husband's affect of immovability.

"The real question," Shea intones, "is whether the system is so out of whack that it can ever be restored."

The beach is on Grand Isle, the biggest barrier island off the Louisiana coast and the only one remaining on which people dare to live. It is no more than an oblong blob of sand, half a mile wide and seven or so long. Its highest point is a dune twelve feet above sea level, running the island's length back of the beach along the Gulf. The Army Corps of Engineers built it to protect mostly weekend or summer camps that are just inland—places with names like Banana

Beach, U're in Heaven, Charlie's Dream, and Fish Tales. Here, where the camps sit vulnerable against the Gulf, the island is no more than two feet above sea level. Despite the protective dune, virtually every building on the island is up on stilts, a precaution against the inevitable. In fact, when you travel the length of Grand Isle, the predominant impression is of thousands of erect wooden legs.

With steady solemnity, Shea announces that we are standing on "the fastest disappearing land in the United States—a thirty-foot width of beach vanishes every year." Not to worry; the United States Geological Survey projects, with a startling sense for exactitude, that the island will be around until the year 2948. Many people are not so optimistic, however, saying that by 2050 it will be awash unless something is done. As expansively far apart as the two dates are, the message is clear: at some time, Grand Isle, like all barrier islands from Cape Cod's Monomoy to Texas's Padre Island, is scheduled to disappear. The last of their sands will sweep east or west, north or south, in the same cyclical journey that the remains of all barrier islands take—upheaved, swirled away, dumped in vast underwater shoals, and then hurled off again in a slow motion at the whim of current, tide, wind, and time. The consequence of this diaspora off Louisiana is perhaps more forbidding than elsewhere. After the islands are gone off this coast, the soft, low marshes are next. But they are without defenses. They are so much fodder to the Gulf's winds and waves.

Shea, a professor of coastal geology and geophysics at the University of New Orleans, often brings his classes to this beach to impart three lessons, ones so important that they should be shared with an audience far greater than his handful of students. The first is that barrier islands off the Louisiana coast—actually sand-bone remnants of the Mississippi's abandoned deltas—are as delicate as pearl necklaces. Compassionless winds of a winter storm can plow hundreds of water-driven channels over and through their gentle beaches within hours, fragmenting them into stricken islets and helpless outposts of sand, shell, and matted vegetation, to be pulverized into oblivion by the next wind. Shea rarely travels without a series of satellite photographs of the Chandeleur Islands, a once legendary necklace that still decorates the river's present long neck into the Gulf.

The Chandeleurs are the last remnant of one of the Mississippi's earliest deltas, this one built up layer by layer around seven thousand years ago. He displays the photographs like a courtroom lawyer laying out for jurors a sequence of actions that has resulted in the destruction of something once held in awe. In 1998, Hurricane Georges grabbed those beautiful Chandeleur beaches with the little dunes overlooking their curves and shattered them into a thousand shards. The storm surges breached channels across the sands and plowed up marsh grass. The winds sent currents careening into Breton Sound, one of the world's most productive shrimping and oystering grounds. The hurricane destroyed without mercy, its pitiless pounding of nature leaving people confused and sad, and birds, fish, mangroves, and shimmering grasses forced to find a new life in an upheaved world, or to die.

Shea's second lesson is about the consequences of nature's, or man's, disregard. The humble stretches of sand that are barrier islands form a crucial phalanx of security for the nearby marshy mainland and for the estuaries they form against the mainland. In these estuaries, the riches of Louisiana thrive, not just oil and gas but oysters, shrimp, and the fifty other shellfish and finfish species that inhabit the Gulf. Without the barriers, storms have free reign to blast in from the Gulf's open waters and strike the mainland with uncompromised fury. Without them, the waters against the coast will turn into the same saltiness of the Gulf, too much in particular for oysters. Waters greater than 50 percent the salinity of the Gulf can spell death for oysters. They will fall victim to the black drum and the oyster drill.

Here's Shea's third message to his students, who are all in training to be in the business of rebuilding this land. It is his most important. It is that although nature—with a lot of help from people—is destroying these islands, they can be fixed. At least that is what Shea hopes. Fixing is part of what Shea does, besides teaching and research. He hires himself out as a consultant to organizations and people with interests in the coast: the Army Corps of Engineers, the Environmental Protection Agency (EPA), United States Geological Survey, the National Marine Fisheries Service, lawyers suing oil companies, oil and pipeline companies with a spill to clean up—companies and people accustomed to spending millions of dollars trying to repair years of ambivalence, ignorance, and stupidity, much of it their own. He calls his ideas for fixing "K-Mart science," likening it to perusing shelves of coastal research for the right repair tool. Some of what he finds on the shelves was put there through his own research on such locally important issues as sediment movement and densities, barrier island migration, under-water sand-shoal formation, and local sea-level rise.

He is also a partner in an oil spill response company, providing technical advice when a pipe leak fouls a portion of the river or coast-line. Given the amount of oil and gas production here, one would think he would be on constant duty. Yet big spills, of the *Exxon Valdez* caliber, are surprisingly infrequent here. Small ones—one hundred barrels here and there from a burst pipeline—are more typical but when a call comes in, the height of the remuneration he receives for his advice just dumbfounds him.

One thing he tells his students, and his clients, is that barrier islands do not have to disappear at their present rate—that their disintegration can be slowed down, but only if people learn to respect the natural processes that create, and are bound eventually to destroy, the islands. And there is something else he tells them— deeper, disquieting, and challenging—that no one quite understands how these islands behave—that the influences constantly at work on them, while almost seeming to be comprehensible, have a mind of their own in the end. To think that the islands' movements can be controlled merely by throwing lines of boulders in front of waves and wind or jetties piercing the near shore currents is an act of pure arrogance. In reality, these islands are only reluctantly amenable to being propped up, rather than fixed. In a profound sense, this misunderstanding typifies the way people have dealt over the years with this unique land.

Most coastlines are contrary, at least to the designs of human beings. They move, except for those formed of bedrock, in frustrating ways that ignore homes nestled atop dunes or fastened to seaside cliffs. The fate of the landmark Cape Hatteras lighthouse stands out in recent years as exemplifying the force of coastal erosion. When built in 1870, the candy-cane-striped monument rose 1,500 feet inland from the shore. But a century of pounding surf and hurricanes ate the beach back to the lighthouse's foundations. A massive and successful effort to save the historic site resulted in moving it a third of a mile back from the surf crashing on the encroaching coast.

Efforts to stabilize moving shorelines have proven as frustrating as the movements themselves. Particularly, some measures tend to result in an increasing lack of stability. Seawalls built to protect coasts, for example, compromise offshore drainage into oceans, drainage that contains the very sediments that nourish beaches. Erosion is also exacerbated when waves smashing into seawalls send the force of their water down onto a beach and chew it away. While jetties, groins, and breakwaters tend to preserve, and even build up, beaches on their up-current sides, the down-current sides receive no material and erode away. When it comes to saving beaches, there are "winners and losers," says Paul Komar, an oceanographer at Oregon State University who has investigated coastal erosion around the world. There are no all-winners.

As Shea talks about preserving the barrier islands, particularly when he is standing on the beach of Grand Isle, his voice edges up and then verges on frenzy. With good reason, as it turns out. He is concerned, as everyone should be—but the national lethargy is startling—about the risk to the maze of oil and gas pipes almost literally under our feet that transports natural gas and crude oil from the offshore platforms in the Gulf of Mexico to refineries along the Mississippi River. The beach is vanishing. The pipes are right beneath it—ready to create the most incredible mess if they were to break under the taunts of a storm. Besides the giant forty-eight-inch LOOP pipeline, eight others cross Grand Isle; three twenty-inch lines carry gas and five smaller ones, crude oil.[1]

---

[1] Over 20,000 miles of oil and gas pipelines crisscross the floor of the Gulf of Mexico off Louisiana, most of them coming ashore at some point and traveling through the marsh.

"The loss of land going on right here is threatening the entire infrastructure of this country," states Shea rather calmly now, given the import of the statement. If the pipes were to be exposed, as they easily might be under merely a slow-moving force-three hurricane (with sustained winds from 110 to 130 miles per hour), the damage could be extraordinary. In fact, the damage has begun. In 1992, Hurricane Andrew whipped across Florida, arced the Gulf, and broadsided the Louisiana coast, tearing up swatches of beach. In a few places, pipelines were hurled up and tied into a "crazy spider's web—a real mess, it was," Windell Curole, general manager of the South Lafourche Levee Disrict, told me. In fact, almost fifty oil and gas platforms were damaged. All were located in the lee of barrier islands, making it a given that as the barrier islands continue to disappear, the onshore oil and gas infrastructure—which accounts for over three thousand wells—becomes more vulnerable.

Some of the pipelines that carry gas and oil in from the Gulf cross a portion of the 35,000 acres of vulnerable marsh that Shea's wife, Cathy, manages for the Edward Wisner Donation in New Orleans. The land is the meager remains of a 1,350,000-acre marsh empire cobbled together on the cheap by Edward Wisner in the 1900s. Wisner grew up in a farming family in Athens, Michigan, but his only interest in farming was to wonder where all the topsoil went every spring. When he was a young man, he followed it south down the Mississippi and quickly came to understand that his family's holdings ended up in the Louisiana marsh. Here must be the richest farmland in all America, where the topsoil from over forty states and southern Canada came to rest whenever the Mississippi overflowed its banks. But first it had to be drained. Wisner began buying the marsh, a place that most people back then saw as wasteland, for between 12 cents and $7.50 per acre. Then he hired hundreds of people to hew drainage canals through it, and to build levees around sections of it. He canvassed small towns through the Midwest, handing out flyers to dirt-poor farmers advertising the country's undiscovered agricultural wonderland. They came along with land-hungry entrepreneurs, convinced that Wisner was a visionary.

The scheme went bust. The natural forces that had created the marsh would not allow such a transformation. The results of Wisner's leveeing are still visible, especially from the air, great square lakes in the middle of the marsh. Some of the marsh had been drained, at least for a while; the levees snaked around their edges keeping out the water. But then the land had subsided; the levees blocked accumulating rainwater and underground seepage from escape. What remains now are geometric water sculptures in the land, like the ones I viewed from the helicopter, complete with drowned farm equipment and buildings. Wisner made a fortune, however, even before the farming bubble burst. Oil and speculators made him rich. Land prices soared after oil was discovered early in the twentieth century, first in Texas and then in coastal Louisiana. Wisner cashed in through outright sales and leases.

When he died in 1915, he left his remaining 53,000 acres—or so it was thought to be—to the city of New Orleans, Charity Hospital in New Orleans, Tulane University, and the Salvation Army, each to benefit from income from the land, along with his heirs. He had also wanted to set up a home for newspaper delivery boys but the lawyers discouraged that. (A later survey discovered that the bequest actually amounted to 40,000 acres.) It is this land, and the distribution of its income—derived from oil and gas royalties, pipeline right-of-ways, fishing, hunting and trapping leases, and 300 camp leases—that is Cathy's domain.

Her job becomes increasingly complicated as the land disappears. For example, she has to fight battles with the state in court over land rights. Under Louisiana law, land that sinks beneath the Gulf or is gobbled up by the Gulf, so that it is covered by water and can be interpreted as being navigable, becomes the state's property. This has huge financial implications. It means that if oil and gas are ever discovered on Wisner property that has sunk, or any other private land, the state is the beneficiary. So much land has disappeared by now that, actually, the state has no idea how much it could lay claim to.

Cathy has to settle more complicated issues, too. Until recently, oil companies were permitted to dump the residue of their drilling into jerry-built dumps next to rigs, levees but no linings required. All sorts of horrors went into them—brine solutions, drilling mud containing barium, and what are called NORMs—naturally occurring radioactive materials—which can be dangerous in concentrated form. Cathy has to make sure that oil companies operating on Wisner land, or seismic surveyors who try to locate oil, don't damage the property. Sometimes in these situations, Cathy and Shea find themselves in an interesting relationship. Sometimes Cathy hires her husband to testify as an expert witness. Sometimes Shea and Cathy find themselves in lawyers' offices on opposite sides of the table.

Shea sweeps his hand over Marchand Bay and settles a finger on an oil platform out in the water. "That was beach twenty years ago," he pronounces, "out there where that rig is."[2] The platform is a half mile offshore. It's an amazing scene in the bay. Oil platforms, oil and gas Christmas trees, and other mechanical, metal-strutted and elephant-legged objects march in odd formation into the haze. In contrast to the soft sand beneath us and the bay's gently ruffled waters, this world of hard angles probing up from the depths seems to defy natural order and organization. But then a fleet of helicopters materializes over the horizon. Soundless, like bumblebees in a larger-than-life, Erector-set flower garden, they separate, each tilting toward a different platform to hover for a second above its metallic foliage before swooping down and disappearing. Emerging moments later, they buzz off to the next platform in unconscious imitation of nature's rhythms.

A couple of shrimpers are out there, too, inching between two platforms, their trawling arms outstretched like soaring wings, a frenzy of gulls arcing and plummeting over their wakes. The mechanical bumblebees imitating real ones, and the gulls adapting to the mechanical harvest beneath the waters, somehow reassures me. It lessens the emotional distance between undulating waves and riveted

[2] A rig is a colloquialism for a drilling derrick, employed when a well is being drilled. A platform is a production facility stationed, in this case, over the water. It controls production from a number of wells that have been drilled in the immediate vicinity, separates oil and gas and water, and controls the flow of gas and oil to the mainland.

strutwork, between netted shrimp and iron-legged islands. If nothing more, the concert of the two before my eyes is a symbol of South Louisiana.

Shea and Cathy don't pay that much attention to what is going on out in the bay. What is novel to them at the moment is my fascination. They tell me I should witness the scene at night from the LUMCON (Louisiana Universities Marine Consortium) observation tower in Cocodrie across Timbalier Bay. "It's like a city out there in the Gulf," Cathy says with a note of pride. "It's the twenty-first century going out two hundred miles." She exaggerates. By the middle of the twenty-first century, there won't be any technology out there among the gulls and shrimp. The gas and oil will long since have been pumped away, the city lights in the Gulf extinguished.

Shea turns back to the sand and now he squats down and begins to draw in it with his fingers, in preparation for another lesson. He traces the outline of the Grand Isle beach—a straight line and then a quick curve inland, then a return to the imaginary shoreline which carries on parallel to the coast and a buffer from the Gulf. He stands, his face the color of cooked shrimp, contrasting against the white spikes of an informally trimmed beard. He angles his sneaker at the indentation and points out where we are, tapping at the sand. I look up and down the beach and see that right where we are standing, the beach takes an abrupt inland jog before resuming its line against the Gulf.

He launches into a gesticulating body-lurching, sand-scuffling monologue on barrier island mismanagement. Powerful Gulf currents that sweep along the coast here are stripping this part of the island of its sand skeleton, depositing it over time to a great underwater sepulchre to the east, he explains. At some point, other currents or perhaps a storm will begin harvesting this lode and create a new island or add to an existing one. Notoriously plastic, barrier islands are in the habit of changing shape, disappearing, reappearing, breaking up into sections and rejoining over time. Coastal geologists call them "ephemeral islands."

Despite the behavior being intrinsic to the ongoing process of beach destruction and beach rebirth, it is not pleasing to the human eye. People are put off by such seeming unruliness of nature; it defies their sense of orderliness and continuity. That, and the fact that you can't swim when your beach has been carried off. So the people of Grand Isle set about to get their beach back, especially important to the weekenders who swell the year-round population of three thousand oil workers and fishermen to ten thousand. A decision was made some years ago to fight nature, as coastal communities in many places have done. Grand Isle planned to replenish its beach by bringing in sand dredged up from the shallows just off the coast. It seemed such a logical idea. The state provided the money, the Army Corps of Engineers, the permit. A dredging company was hired and got to work pumping up sand through a big pipe and spreading it along the length of the beach. If anyone thought it peculiar that the dredge worked just offshore from where Shea is now discoursing, nothing was said. The tons of sand that the dredge sucked up was re-creating the beach, after all. Nothing was said until people noticed something odd. The beach nearest to the dredge was disappearing in a neat semicircle, just the way Shea had diagrammed it on the beach. Suddenly, a dawning occurred. The deep hole offshore which the dredge had gouged out was gobbling sand right off the nearest beach, like a mysterious maw that could not be satiated.

Shea shakes his head in exasperation. Though the dredging was brought to a halt, the beach kept disappearing as though a cookie cutter from above was bent on creating an ever larger treat. Shea sees symbolism here. "We tend to want to use static solutions to fix dynamic situations," he lectures. By "dynamic," he means that once the hole was made, it just got bigger and bigger through a combination of physical forces set in motion by the dredging, pulling more and more beach into the hole.

Time to try another solution; unfortunately, it was another static one. The then-mayor of Grand Isle, Shea continues—looking more and more like the preacher on the beach exhorting against the sins of coastal land management—came up with the idea of erecting a rock jetty perpendicular to the beach just down current from the growing hole. The mayor said the jetty would be a "sand-making machine." There are no mountain ranges here in South Louisiana, not even rock formations, no source of boulders to go into jetties. But help was nearby. A politician who happened to own a granite quarry in Kentucky offered his services. Tons of rock made its way by truck and barge to Grand Isle. The idea should have worked, but it didn't—proof that human efforts to counteract nature often underestimate nature's ways. More jetties were the answer, the mayor counseled his constituents. In the end, the town spent $12 million on rock to make a series of jetties whose remains today look like rough scabs in the sand. They didn't do a whit of good as, in the long run, they haven't in many coastal communities. The down-current side always loses eventually.

Shea jerks his arms up in the air and brings them down sharply, cutting the hot air in a finale. "It's like we are in this idiot loop down here that no one can get out of. It's like that movie, *The Truman Show*. We're in a bubble down here and can't see what's really happening."

Shea gets worked up over things like this because the role of barrier islands as buffers against future storms is not a matter to be taken lightly, especially when almost three-quarters of the state's population live less than a foot above sea level, and a good many of those *below* sea level. A century ago, barrier islands protected coastal residents more effectively than today. They were sturdier and seemingly more permanently planted. Tight-knit farming and fishing communities grew on some of them. Barataria Bay was filled with red-sailed luggers, their decks brimming every evening with baskets of freshly dredged oysters or seined shrimp and fish. On the sandy high ground of Grand Isle, melons, cauliflower, and cucumbers thrived. Produce went north, first by boat and then rail, sprinkling the Mississippi River valley with their seeds. The island also boasted what was called the world's largest turtle farm of diamond-back terrapins. The creatures ended up in stews in restaurants in New York, Philadelphia, and Washington, but were also a staple of ship crews on long voyages. Unlike most sources of live meat, such as chickens or pigs, terrapins could live in a ship's hold for months without food or water, a ready source of protein. At one time, the herd, as it was called, numbered twenty thousand head.

But it was after the Civil War that the islands became known for their salubrious breezes during the hot summers and as escape destinations from steaming and yellow-fever ridden New Orleans. Two retreats went up on Grand Isle, the Kranz Hotel and the Ocean Club. The Kranz, built to remind guests of plantation life, consisted of thirty-eight cottages along a grassy street, the ballroom at one end and the dining room at the other. Tram cars shuttled vacationers to the beach for the designated swimming hours— 5 A.M., noon, and 6 P.M.

The Ocean Club was far more elaborate, built for the vacationing cream of the south with 160 bedrooms facing the beach and a grand dining hall that could seat 250 guests. The developers, knowing that the hotel's siting was precarious, insisted on double-framing the entire structure, which entailed the use of over thirty-four miles of wood merely for the structure's skeleton.

Nature's fury had no regard for such care. In 1888, a storm swept the island, tearing away the Ocean Club's beach houses and forcing the guests to evacuate to nearby Fort Livingston, a massive brick fortress constructed in the 1830s to ward off the British. However, that began to sink into the marsh as soon as it was finished. Five years later, in 1893, Grand Isle as a resort island ended when a powerful hurricane again buried the island in tons of roiling water, this time in the dead of night. The Kranz Hotel was totally destroyed. "A terrible gust of wind struck the house and knocked it over," reported Mr. Kranz in the *Daily Picayune*.

He was pinned under a falling beam and as he felt the water rising around him, his predominant thought was that he would be dead within the hour. Fortunately, the weight bearing down on him rendered him unconscious and he did not have to feel himself drowning. The rising water proved to be his savior. Eventually, it lifted the beam, which floated away from the hotel's sloshing debris. It carried Kranz with it, his hands gripped into its wood like a raptor's talons. When he came to several hours later, he and the beam were safely bobbing out in the Gulf.

And this happened on the biggest of the barrier islands. A smaller, far more fragile spit of land lay just to the west—Cheniere Caminada, a name that befits Louisiana's history because it combines French—*chenière*, meaning oak tree—with Spanish—*caminada*, meaning walkway. Beginning in the 1830s, a wondrous potpourri of around 1,500 Yugoslavians, Chinese, Italians, Malays, and free Blacks lived here in little palmetto houses. The place was known for its Saturday night dances, which cost twenty-five cents admission.

Here's a description of the storm from George Reed, resident of Cheniere Caminada for nineteen years:

> I resided in a small house at the end of the island, with my wife and six children, all of whom were the victims of wind and wave. My house and all I owned on earth have been swept away. It was about five o'clock on the evening of Sunday last (Sunday, Sept. 31, 1893) when the storm commenced. At nine o'clock the water had risen to a depth of several feet over the banks of the island.
>
> The wind increased in velocity, and shook from the foundations the frail house in which both I and my wife were sheltered. I realized that to remain there meant certain death to all of us. I left the house with my wife and children, and after battling with the weather for several minutes, we found a cabin which we entered.
>
> Several families had already left or were driven from their homes and were seeking shelter at this cabin. I counted, including myself, wife and children, forty-one persons in all, distributed in the few small rooms. After remaining at the latter place for about an hour, a violent gust of wind struck the building and obliged us all to hasten to the outside. The house was dashed to pieces, and out of the forty-one souls that it had recently sheltered, only twenty-one were living, the balance having been thrown out to the bay and drowned. I looked around and found that my wife and children were yet alive. Together we again attempted to find a place of safety.

Only four homes still stood after the 1893 hurricane. Eight hundred residents died. Seventy-eight people sought refuge in one house; the roof collapsed, killing seventy-four of them. The dead began to smell so much that they were buried without coffins.

Today, you can still see the twisted foundations of some of their homes set among the skeletons of live oaks quivering in the wind off the Gulf.

The survivors moved north up Bayou Lafourche, where they live today in the towns of Golden Meadow, Cut Off, and Leesville in the uneasy knowledge that a hurricane next year or the year after could drown them with the same ease it destroyed their forebears. Windell Curole, the South Lafourche Levee District manager, is a descendant. Appropriately, given his family history, he is largely responsible for the forty-four miles of levees that enclose Golden Meadow and the six pumps in case the levees are breached or overtopped. The land inside the levees is referred to, with some engineer's nod toward the poetic, as fastlands—held fast by the levees, but subsiding nevertheless.

Curole's grandfather was six months old when the 1893 hurricane struck Cheniere Caminada. The family story has it that during the hurricane, the infant's parents climbed the stairs in their house to the attic to escape the rising water. In the confusion, the baby slipped and fell into the pitch-black swirling water below. Here's the miracle: his panic-stricken parents groped down into the darkness, flailing to feel something. They did, a soggy bundle, alive and squalling.

Windell says quite simply, but startlingly, "If they hadn't, I would not be here talking to you." Hurricanes are indiscriminate. They are also certain.

A month after witnessing Shea's college lecture and impromptu sermon on the Grand Isle beach, I am onboard the *Mud Lump*, clinging to anything I can and tightening my stomach muscles as the hull ricochets off Gulf swells. It's an open aluminum boat with a little enclosure built around the wheel amidships and its drumbeat across the water on this brilliant May morning makes me think of jackhammers on a hot pavement. Shea is driving and he looks like he knows what he's doing, leaning into the wheel as the boat arcs and leaps, and swerving its prow away from dolphins and

occasional sea turtles, but mostly from gas and oil industry paraphernalia.

He must be reminded of his youth right now, of growing up in
Jacksonville, Florida, where he was "a tidal-inlet kind of guy," as he
once explained to me. I picture him as a kid wading the shallows,
minnow net and fishing pole in hand. As a teenager, he said, "I
fished and surfed my way through Jacksonville University," not
thinking a hoot about the future. He worked summers and vaca-
tions guiding sports fishermen around estuaries and inlets not
much different from those of the Louisiana coast. Then his boss
offered to make him a captain of a commercial fishing boat—a real
job that carried responsibility for real assets. Shea was sorely
tempted. But such a position was burdened by other responsibili-
ties—more cultural obligations, really. Boozing was one.
Smuggling cocaine and doing cocaine were also part of the job.

Shea began thinking about the consequences, began noticing
what happened to some of the people he knew in the business. His
father, an "infamous trial lawyer, a legend in his own time," as Shea
refers to him, helped him think. Perry Penland, Sr., Shea told me
on more than one occasion, with a lilt of pride raising his voice, is a
formidable and colorful lawyer in Florida. Still practicing, he
served as a defense lawyer for Lester Maddox and George Wallace,
as well as the quirk of representing songwriters for Elvis Presley.
He also carried a .38 revolver, which he called "the owl's head."
Shea remembers him pulling it out of his shoulder holster one
night in their driveway and firing off a round at a figure fleeing
across the yard. His father told Shea, rather vaguely, that it was an
FBI agent on surveillance duty.

Despite the famous clients, Perry Penland's real bread and butter has always been, and continues to be, plaintiff law, representing
car-crash survivors, job-accident victims—victims of any kind.
When Shea was a kid, his father piled him and his younger brother
into his big Cadillac on weekends and drove them to hospitals and
homes where these victims were recuperating, or merely surviving.
"He wanted us to see people who had been maimed in auto accidents so we would know how lucky we were," Shea remembers.

Floundering about trying to find a clean occupation, trying to please his father, trying never to leave the water, Shea eventually hit on coastal geomorphology and entered a masters degree program at LSU in 1977. It was the beginning of a bunch of bad luck. In the summer of 1980, he found himself in Germany on the Elbe River delta and East Friesian Islands, studying geological stability. The U.S. Navy funded the research. Only later did he learn that the Navy was trying to determine the best kind of amphibious landing craft to use for military operations. Shea never could find out why the Navy was interested in this particular location; that part of the project was classified. And before he could finish his work, the Navy withdrew the funding—every graduate student's nightmare.

Befuddled, he returned to LSU with only half his research completed. He managed to survive academically, working on a university barrier island project, but the funding for that ran out. Shea headed off to British Columbia as an environmental consultant for oil and gas development. Then the price of oil dropped and the hard-luck kid was out of a job.

The bad times didn't end until Louisiana realized that its southern edge was disappearing. Suddenly, LSU found itself with all sorts of money for coastal research. Shea picked up his graduate research and found that he could pluck grant money off trees. Finally, he was on the road to becoming, as his wife, Cathy, now proudly calls him, an "eminent scientist" with a doctorate in coastal geomorphology.

He looks astern of the *Mud Lump* every few minutes to check the progress of the cigar boat following us. The boat used to run drugs until the State Department of Wildlife and Fisheries confiscated it. Staffers now use it for any number of missions—busting poachers, running down more drug runners, or today, for assessing the extent of one of coastal Louisiana's most obvious weaknesses—its barrier islands—and trying to find a solution to their habit of melting away under a storm's wrath.

Aboard the two boats and a third one trailing the cigar boat—a little tubby thing that looks like a poor imitation of a New

England lobster boat—are representatives of federal and state agencies, all CWPPRA task force agents whose mission today is to assess proposals put forth to save the barrier islands. The federal people are from the Environmental Protection Agency (EPA), the National Marine Fisheries Service (NMFS), U.S. Fish and Wildlife Service (USWFS), the Army Corps of Engineers, and the Natural Resources Conservation Service (NRCS), a branch of the Department of Agriculture. State people include staffers from Wildlife and Fisheries and from the Department of Natural Resources. There's a gaggle of reps from NRCS in the wallowing wannabe lobster boat. They are dressed in jeans supported by wide belts with big buckles. They look like they should be riding a tractor rather than tossing around in the Gulf and, indeed, their allegiance has long been to farmers rather than fishermen.

The composition of this group as a whole—many of its members engineers and biologists without PhDs—is remarkable. Given the traditional divisiveness endemic in this state—economic and political backbiting which has done so much to stultify environmental concerns—this group of thirteen people display relative unity. Traditionally, each state and federal agency has its particular conservation, or land use, ethic—its way of being. Each favors management techniques that please its constituents. NRCS has always liked to build earthen levees, for example, around farms in danger of sinking into the marsh. The Army Corps of Engineers, along with erecting levees on the Mississippi and elsewhere, likes to halt erosive forces by erecting rock jetties as protective barriers. The Corps has a reputation for concrete, rock, big equipment, and big money. It prefers big projects such as the canals built to drain the Florida Everglades by sending 1.7 billion gallons of fresh water into the Atlantic every day instead of through the sawgrass. And National Marine Fisheries Service (NMFS), a branch of the Department of Commerce, likes to plant marsh grasses and build dunes to enhance production of commercial aquatic life. Separate philosophies operating within tight budgets long resulted in spats over funds, turf, and popularity. The marsh lost.

Since 1990, with the passage of the Breaux Act, much of the

competition, but not all, has diminished and what I am seeing, or supposed to see, as we slam across the Gulf, is the new spirit of cooperation codified by the Act. The passage of CWPPRA brought a hope of relief to those who care about the fate of South Louisiana wetlands, which does not include everyone, particularly those in New Orleans and environs, where half the state's population lives. Nevertheless, the Act has at last reined the snarlingly independent and largely ineffective conservation agencies into some sort of unity that focuses on the welfare of the coast. At least that is the way it's supposed to work, but traditions die hard.

We are headed to Timbalier Island, one of the remains of the low-lying sand islands that used to rim the Louisiana coast. It's out there somewhere in the Gulf. Oil industry pilings angle up from the roiling water ahead, at times singly, or in tight-packed clusters, or in neat lines, as if the gods had dropped a bunch of massive Pick Up Sticks from the thunderheads that perpetually dwell on the horizon. Pipes budding grotesque valves and gauges jut up from the water. Nervous pelicans and terns perch atop them. Flames shoot up from thin stacks off production platforms hulking ahead of us, fearsome metallic skeletal monsters that in a more pristine setting would have no business among wheeling birds and nature's contours. Oddly, the clash is minimal; soft wings and hard steel complement each other here. How comfortably the extravagant bird life takes to mankind's mess. How strange it would be for the birds if the Gulf were swept clean of their ugly perches. In fact, the struts of these monsters save the lives of untold neotropical birds on spring migrations to the mainland U.S. from their wintering grounds in Central and South America. Oil workers frequently find exhausted warblers tottering on them, recouping the last of their energy after the four-hundred-mile flight from the Yucatan, before continuing the few miles to the mainland.

Shea and everyone else on board seems to know where the island is, though I see nothing on the horizon but the platforms. The people in the boats have great and little plans to tinker this barrier island and others back to life. There is a palpable sense of urgency here to do this, a hospital-emergency-room need to save a life. Given most of the islands' present state—rising only five feet or so from the water's surface—I am not sure if the effort is worthwhile. But evidence exists that these islands, even their remnants, serve as buffers against storms barreling in from the Gulf; this is reason enough to tackle the case.

Yet there appears to be a suggestion of a dearth of scientific expertise aboard. Pat Kelly stands swaying next to me, a young man from the NMFS. His posture is marine-like, stiff, somber, disciplined. I look for humor around the edges but it isn't there. Then I try to yell some questions about coastal erosion over the roar of the outboard and the whack-whack of the hull on the waves, but the response is minimal. Maybe he's just shy of me, the writer from New York.

Another young man named Troy has a similar bearing; he even has camouflage pants on. He's from the EPA office in Dallas. Up on the bow is Tim Landers, also from EPA. He looks about sixteen and there's not an ounce of fat on him. He looks so different from most people in this part of the country, like a dancer, that I find myself transfixed. These are the young turks of coastal salvation, at least I suppose they are, but not grounded in science. Mark Hester is older and, like Shea, a real scientist. He's a botanist and has pulled his white socks up high on his white calves, giving him a nerdy look. I will discover an absolute joy in him over the plants on the islands as he lovingly talks to me about their traits. Then there's Gary Rauber, and he's a kick from the Army Corps of Engineers, a maverick with a wicked eye and sharp tongue whom I will find to be delightedly disdainful of everything going on around him this day. Sue Hawes is with us, too, from the Corps, backpack, binoculars, and bird identification book at the ready.

Timbalier, so low on the water that it seems just to materialize out of the swells, turns out to be as scrawny a neck of sand as any of the other barrier islands. Research material that I had collected while still in New York makes mention of the Gulf barrier islands' sand dunes as offering protection against storms and hurricanes. I envisioned the rolling sand hills of Cape Cod or eastern Long Island. But here in Louisiana, what is called a sand dune averages only three feet high, disappointing to a Yankee but understandable given South Louisiana's flatness, where any elevation calls for comment. As far as providing protection, I don't know, but Louisianans seem inordinately proud of their little dunes, regarding them as making the difference between survival of the marsh to the north and destruction.

We are not talking much island here—half a mile long parallel to the coast and a couple of hundred yards wide. Marsh hay, beach tea, and groundsel—a four-foot-high bush that, given its proliferation, has obviously mastered survival in a merciless environment—blanket the island's inland length. A sand beach ribbons the Gulf side, barren and empty of protective vegetation but well stocked with plastic debris. Here is where barrier islands can die, their fragile beaches pierced by waves and wind until they are carved up under nature's knife and left in bits and pieces to be worked down to nothing by succeeding storms.

Shell Oil holds the drilling lease to Timbalier. It advertises its dominion with a scattering of large NO TREPASSING—PRIVATE PROPERTY—TRESPASSERS WILL BE PROSECUTED, etc., signs on pilings at the entrance of a stubby canal dredged to give workboats access. No one in Louisiana pays any attention to such orders, which mostly serve to discourage people from suing if they break a leg on private property.

The boats tie up to a rickety wharf and we unload and file with great solemnity along a little path to the Gulf side. We are headed to a place that recent storms have breached and which future ones are sure to exploit, cutting the island in half and opening a passage for high water to invade the marsh on the mainland. The consequences are greatly feared. The invasion of wave after wave can tear the marsh to tatters, pushing it further and further inland, killing off whatever ability it still has to protect the farms and towns just to the north. Of more immediate consequence, saltier water flooding over the marsh, probing rivulet channels previously occupied by freshwater, can help decimate the estuaries that countless baby shrimp depend on for sustenance. The economic consequences to the Gulf shrimping fleet, already in a losing race against

Asian and South American shrimp farms, is as black as the fiercest storm bearing down on the coast.

The EPA came up with a plan to repair the damage. Tim Landers is point man today. He has to sell the plan to the rest of the group who, after all the proposed projects have been viewed, assessed, and ranked, will choose around thirty to be funded. This is where the CWPPRA process, as it is called, shows some glitches. Those agencies which have sponsored projects that are awarded funding receive a lot of money. Complaints are frequent that one agency or another has sponsored a disproportionate number of projects, or assessed them to favor one agency over another. And there's some backroom horse trading that goes on, too. At this writing, NMFS has won the jackpot with over sixty projects approved, funded for tens of millions of dollars.

As we file along the narrow path weaving through the cord-grass to the beach, royal terns with their stunning orange beaks hover above us. Gigantic cumulus clouds give a profound holiness to the horizon. I feel like the mood should be lighter. I am put in mind of rambles to unknown beaches for a swim, shells to be discovered, squeals of delighted children, sand-textured egg salad sandwiches. Not here, not now.

There's a subtlety in this march to fix nature. Those who are really interested in this particular repair work lead the way. That means Tim Landers and his EPA partner, Troy, whose camouflage pants weigh me down as much as they must him. They look so incongruous here on the beach under the blazing sun. Following close on Landers's EPA heels is the NMFS, then the USWFS, then the Corps. That includes Sue Hawes and Gary Rauber. Sue zigzags along the course, poking through bits of nature as she goes—shells, feathers, and lonely plants—seeming to be endlessly intrigued by whatever appears in front of her deeply tanned face framed by gray curls. Gary follows, by himself, looking worried and nervous, like someone who has accidentally wandered into a threatening neighborhood. Bringing up the definite rear, so far behind that we lose sight of it is the NRCS, cowboys who have shown only minor interest in saving a pile of sand. Shea is up toward the front, hefting both an aloof authority and authority-bearing rolls of satellite photos of the area. He's here as a consultant to the EPA and the Corps, an expert witness (which he does a lot of in court cases) to hear the "remediation plans," as they are called, and to offer criticisms and suggestions. He earns a good day's wage for this kind of work. "I am a gold mine to these people," he tells me later. "What I do is just what they need, and this kind of work offers a lot of low-hanging fruit."

A half mile of trudging along the beach, kicking plastic containers out of the way, and marveling at some substantial redfish rotting on the sand, Landers stops us. Like a ratty sandlot stickball team, we gather around him. Even his yellow-tinted sunglasses cannot hide his nervousness. He's developing a sales pitch but he does not look like a salesman. The muscles of his lean jaw work up and down like he is chewing on something tough. Where we are gathered is a particularly low section of the island, in fact, only a foot or so above sea level. From here the island descends in the direction of the mainland marsh across the pass, entering the water on the inland side as a soggy sandbar. It's obvious that any storm worth its salt would careen water right over the island at this point. Any sustained winds would carve a channel right where we are standing.

Landers begins his presentation by telling us that Hurricane Andrew in 1992, and subsequent storms, did just those things. Though sediments carried in the westward-flowing currents have filled the channel in recently, the EPA thinks it can do better than nature. In fact, the EPA can stop nature. Landers proposes barging in hundreds of tons of sand to create a five-foot-high sand dune two hundred feet wide. He "guesstimates" that the work will result in the creation of almost seventy acres of marsh. The audience— the judges—begin asking questions, all except the cowboys who have straggled up late and hover on the circle's periphery, swatting insects and looking bored. Someone wonders about the merits of bringing in "hard structures," meaning snow fences, from the north. Erected at right angles to the prevailing easterly winds, they have effectively produced sand drifts against their windward sides. Unfortunately, the drifts are not oriented to buffer storms pounding in from the Gulf. Where will the sand come from, someone asks? And if it is dredged from one area and creates a hole offshore, will that be tantamount to creating a "wave-focusing event"? The terms "updrift" and "downdrift" are bandied about together with "downdrift deficit."

Of all the terms, "guesstimate" intrigues me the most. During a three-day period, visiting a handful of islands with these folks to hear how they are going to save them, the uncertainty expressed by that oft-repeated fabricated phrase is downright alarming. On the other hand, such talk hedges against the possibility of fallibility, anathema to the engineering mentality that tends toward linear thought and predicted outcomes. So "guesstimates," which I gradually learn can be appended to anything engineers in this part of the world tackle, is really a way of saying that actually no one knows what nature has in store.

Shea has been silent, listening to the presentation and to the questions, again planted in the sand, as if he had grown out of it. Toward what he perceives to be the end of the sales pitch, he rolls out a photograph on the sand and weights down the corners with some rocks. Then he stands up and listens some more. At a moment of silence, he quietly utters a startling truth, given the proposed solution we have just heard, albeit fudged by "guesstimates": "You can't stop the Gulf from coming in here if it wants to," he intones with awful certitude. The effect of this pronouncement by the expert consultant is that kind of quiet you hear during a play when an actor has forgotten lines and everyone else on stage is suspended in unnatural poses. Shea tilts his bearded face down at his big photo on the sand and everyone else looks down, too, and Shea points to where we are and we see how narrow the island is at this point. He suggests that the EPA think of building a marsh behind the proposed dune, accomplished by barging in about a billion more buckets of sand, and arranging it so it rests low on the water's surface. He calls it a "marsh platform." The roots of the vegetation sure to sprout from it would act as an anchor against storm ravages.

"Hurricane Andrew taught us that wider and higher islands are better off," he states, and then adds that island width has a lot to do with maintaining what he terms "island integrity." His listeners are silent, but wavering eyes indicate that some are beginning to let their thoughts meander. He waits for eyes to return to his eyes. When all is still, he introduces a concept that is probably new to most people here. It's called the paralic zone, a geologic phrase of, literally, earth-moving hugeness. In concept, it is this: the Gulf coast is shuffling northward inch by inch as the land subsides, a geologic process that rising sea levels are hastening. Barrier islands, or what is left of them, are also inching northward as the platforms they rest on move in concert with the coast. This migration has implications for the long term. As Louisiana worries more about the impact of hurricanes, as it should, plans are brewing to throw up a broad band of levees running more or less parallel to the coast, generally in an east-west direction to keep out the storm surges. But if CWPPRA, in its effort to save barrier islands, erects sand dunes atop their remains, high waters off the Gulf could become entrapped between the lines of levees. The result would cause exponential land loss—expanses of marsh would drown and slough off into the Gulf.

Fascinating in concept, but not about to happen tomorrow. That dune proposed by Landers means money from the feds, about $5 million. Shea's suggested questioning of its value does not cut it with this group. Eyes begin wavering again. Some glaze but most settle on bits of shell or driftwood in the near distance. Feet scuff backward to play footsie with the shell or driftwood. The circle breaks, leaving Shea with an audience that has dwindled to one or two. He stops talking eventually, winding his gesticulations down like the rotors of a just landed helicopter that has cut its engines.

On the way back to the boats, I walk with Sue Hawes and Mark Hester, who have both maintained polite attentiveness during the display of technical terminology. Sue seems enraptured by a royal tern that has decided to accompany us, but stops at something she thinks I should observe. It is the skeleton of a bush, its trunk buried in sand. Against the remains of its limbs, little mountains of sand have accumulated, their peaks no more than six inches high. The effect is demonstrative of the malleability of these islands, wiped clean before the ceaseless wind and then built up again by the same wind. "You see how little it takes to make these islands right again—just a couple of branches have a tremendous effect," she marvels.

Mark approaches nature's little construction site and examines the bush. It's the remains of a black mangrove. When he has made the identification, his eyes soften with delight, an emotion that I will see manifested many times over the next few days as I realize the profundity of this man's affection for plants. A black mangrove, even in skeletal form, is worth an exclamation. Its roots can hold an island together for years. In Louisiana, the plant is restricted to the barrier islands; the mainland marsh just a mile or two away is too northerly for their tenderness. It doesn't take much of a chill to make them fold their leaves and die within hours. If that happens on a barrier island, the island will begin to die as the mangrove roots loosen their grip on the sand, freeing its grains for the wind to carry off. "What's really neat about black mangroves," says Hester, "is that their seeds have already germinated when they drop. They have little roots on them and they begin growing as soon as they hit the ground."

It's refreshing to hear these two talk of nature, unencumbered by manipulations. But such appreciation is not going to save these islands. Neither is engineering.

The Cocodrie Marina is where everyone ends up at night in this part of Louisiana. I go to dinner there with Shea and three or four other people from the trip. The place is varnished and kitschy but feels right; shiny redfish, speckled trout, dolphin, and shark assume dramatic poses from plaques on every wall. Buoys, crab traps, oyster tongs, ship's wheels, the whole nine yards, accompany them. A huge aquarium sits off to one side, empty now but the sand on its floor reminds me of one of the barrier islands.

We are seated at a table eyed by a tarpon staring down at us. Everyone is a mite stiff—men strangers to each other—trying to socialize. Shea's the definite alpha male here, sitting at one end looking authoritative and the other folks looking to him. I ask the waitress, who looks sassy, what kind of beer they have. I haven't been in rural Louisiana long enough recently to know better. I expect her to rattle off a list that these days might include Sam Adams, Guinness, and Dos Equis, or at least Corona. She looks down at me as if she just discovered a stray cat on her living room couch. The agency guys are eyeing me a little strangely and I realize I have set myself another one thousand miles or so apart from them, the first 1,300 being accounted for by the fact that I am from New York. Here, Bud is the beer along with Miller. You can get Heineken, too, if you are willing to be a little different. I tell her Miller and she says she'll try to find one.

The conversation is strange and guarded, a lot of engineering references such as "storm events" and "tidal scours"and "necking down channels" are thrown around. But there is one person, Kevin Roy, from U.S. Fish and Wildlife, who shoots me glances like he is curious about me. In general, the lack of communication—the hastily erected walls—I find intriguing and so I just sit back and watch for a while.

There's considerable long staring at the plastic covered menu which offers everything you might expect in a Louisiana marina restaurant, from crawfish etouffée to fried catfish to oyster po' boys. When the waitress takes our orders, an array of interpersonal relations, perceptions, struggles, and dreams are suddenly added to the menu. Pat opts for fried something, humble and commonplace; Mark Hester, the botanist, hems and haws like he wants something that isn't there, but settles for broiled redfish. Shea chooses grilled shrimp in a wine and cheese sauce, perfunctorily, like he doesn't have to give the matter much thought. It's a fancy dish that sounds showy. The others throw quick glances at Shea as he orders. Shea seems oblivious. Kevin Roy, who had declared his intention to order fried oysters, makes a quick switch to a shrimp dish in some sort of a sauce with a name as sophisticated-sounding as Shea's choice. As Kevin orders, he doesn't appear comfortable; he squirms; he looks up at the waitress as if seeking her approval.

During the meal, Shea obviously wants to get a conversation going. The place is filling up now and he has to talk loud to make himself heard above the din of the sports fishermen. They look well off, big-bellied in logoed, open-throated shirts. Shea asks Pat Kelly why it is that National Marine Fisheries has allowed a marsh alteration to take place that affects fish habitat poorly. Kevin smirks and kids that this kind of inaction is typical of the NMF, but Shea turns to him, a glow growing on his cheeks, and points out that U.S. Fish and Wildlife allowed a culvert installation that drained marshland. Kevin looks down at his shrimp. Pat informs anyone who wants to listen that the Corps should be blamed for the fish habitat alteration. Just then Gary Rauber joins us, Heineken in hand, and totally agrees with Pat. The Corps, his employer, is definitely to blame, he says with a big smile.

Gary looks very content in his naughtiness. He shifts the conversation to the day's discussion of the various plans to save Timbalier. "You guys are just pissing sand in the wind. You are not going to help anything with your little nipples of sand. A force-five hurricane would knock it all out in a second." (This is good traditional Corps thinking, wedded to concrete and rock.) He leans back tilting his chair off the floor and sweeps the beer bottle to his mouth at the same angle as his chair, providing him with a compactness that parallels his statement. Pat and Kevin and Mark chuckle. Gary's an iconoclast; he's glib and smooth, too, and apparently not afraid of offending anyone. So he's fresh air and the others like him for that. And maybe, also, because he might be right.

Chapter Four

<u>Oil Now and Then</u>

First the marsh was mostly empty of people. Fishermen and hunters passed through its bayous on the way to estuaries. Houma Indians occupied some of the bayou banks deep down near the Gulf in Barataria and Terrebonne Bays, driven south over the decades from the high banks of the Mississippi, finding security on the edges of land no one else wanted. Escaped slaves sought refuge in the wetlands, too, living with the Houma. Some Chinese smuggled to New Orleans in barrels and pressed into service drying shrimp lived there, too. Others migrated from California where shrimp drying had become a cottage industry. Eventually, the Chinese settled in villages that rose over the marsh, dominated by stilted drying racks as expansive as two football fields side by side. The settlements were intriguingly named: Bassa Bassa, Cabinash, and the Fifi Islands. At one time, there were seventy-five platform villages scattered around the marsh. Manila Village was the last one, knocked out by Hurricane Betsy in 1965.

Despite these small incursions, the marsh remained untouched in a vastness that people left to its own. The isolation created ideal concealment for pirates. There had to be pirates here where the marsh's mazes could swallow up men and ships with the ease of a fog bank. Jean Lafitte is the most celebrated. Like a heron, he waited quietly in the lee of barrier islands in his nimble schooner. Cumbersome Spanish ships plying the Gulf were his prey. He hefted his contraband, including slaves, to Grand Terre, a barrier island just east of Grand Isle and, along with other pirates, set up a lavish flea market on the beach. New Orleans businessmen sailed to the island on weekends to bargain with this gentleman pirate, who became quite an acceptable, if distant, fixture in New Orleans society. Business grew and Lafitte, emboldened and bolstered by the government's goodwill, bought a warehouse in New Orleans from which he sold his goods, saving his customers an arduous journey. But they say he kept much of his treasure down in the marsh. They say it's buried everywhere, definitely on Grand Isle and on Cheniere Caminada, clear over to Galveston. Of course that's what they say.

In those days, the marsh was not owned by the people. The federal government owned most of it. Then, in 1849 and 1850, high waters barreled down the Mississippi, flooding farms and rampaging through New Orleans. In response, Washington passed the Swamp Lands Act as a mechanism to bail out the flooded areas and keep them dry through levee-building. The Act granted almost twenty-eight million acres of flooded land to the states bordering the Mississippi. After bayou banks were leveed and the marsh drained, states could sell the reclaimed lands—as they were called—continuing the European-induced hamstringing of the river's will to flood its banks. The river rebelled, as it does from time to time. The devastating floods of 1993 along the upper Mississippi are its most recent misbehaviors. The river breached its walls and drowned the reclaimed land—this time with houses, and towns, and farms on it—as it had many times before.

As a result of the Swamp Lands Acts of 1849 and 1850, the edge of the Louisiana marsh changed hands from federal government of state government and then to individual. People began to settle along bayous. The property they purchased was one arpent wide (192 feet) on the bayou by forty deep, going almost 1.5 miles back into the marsh and swamps. They built their houses and planted their gardens on the high ground of the natural levee on the bayou. Cattle and a horse or two were pastured on the land descending from the levee. In back of that was marsh, good for hunting, fishing and trapping, and nothing else. On some early French maps the marsh back there, which then amounted to five- and one-half million acres, was assigned merely: "L'Intérieure Inconnu" or "Territoire aux Sauvages."

Unfit as is, perhaps, but the stuff of dreams. Land grabbers saw paradise. Men like Edward Wisner moved south, buying up huge swaths of marsh for pennies per acre. Wisner's eventual one mil- lion-plus acreage included eleven entire parishes. He set the pace for other land companies—Delacroix, Vermilion, Miami, St. Martin, to name but a few—some with farming ambitions and some with lumbering. The huge cypress trees in the Atchafalaya Swamp began to fall, floated out through still visible canals, and rendered into ribs, planking, joists, and beams for millions of boats and homes. My father, a New England dairy farmer, had a Louisiana cypress water trough for his cows, big as an iron bathtub and far more rot resistant. The cypress silo of his barn never showed a bit of rot, even after a half century of storing wet silage. The big cypress is gone now, except those trees made useless by lightning scars or a tilt or twist to their trunks. The land companies are still there, waiting for the offspring to reach market size. It will be a one-hundred-year wait, at least.

Fur trapping had long been a part-time activity in the Louisiana marshes, etched into sons by fathers down through the generations. Mostly otter, mink, and raccoon and alligators were the quarry. As alligators became rarer, hunters took to burning the marsh to more easily locate 'gator holes. The carbon-enriched land opened a niche for an explosion of three-cornered grass (*Scirpus olneyi* and *S. robustus*), species favored by muskrat. The sudden abundance of the grass, seed pods waving in great formations across the marsh, led, in turn, to an explosion of muskrat. Some trappers did not know what the little animals were when they first saw them. They didn't even have the proper traps for them. But in New York, furriers had long dealt with muskrat from the north.

They knew that a muskrat was actually three animals in one. The back fur is a lustrous black, the sides, a light brown, and the belly, a soft pewter. They knew, too, that the coats, jackets, and hats they made from these three pelts didn't have to be sold as muskrat, a name that lacks a certain something. Far more alluring was to call them French seal or Hudson seal.

At the turn of the twentieth century, a muskrat pelt was worth eight cents, so little that landowners didn't mind trappers taking animals on their land. Then the price began to rise. Trappers, especially those in southeast Louisiana, bore down on the soggy little animals as they realized the beginning of a tremendous bull market in rats. The price jumped to a quarter a pelt, then fifty cents. Landowners and would-be landowners quickly came to understand that the muskrat gave new value to the marsh that had nothing to do with farming it. Gatien Livaudais and his family bought up 12,000 acres of St. Bernard Parish for its muskrat. Where trappers once roamed freely, they were now not only told where to trap but they were forced to sell their furs to land company agents who traveled by boat from one trapper's camp to the next, exacting what some saw as an exorbitant percentage.

Then the oil barons came. The first Europeans to note oil in south Louisiana were members of Hernando De Soto's 1540 expedition down the Mississippi. Explorers then didn't know what the gooey black stuff was. They called it "stone pitch," caulked their boats with it, and tried to wipe their hands. Native Americans poking around the state's peculiar hillocks in the marsh, which are really salt domes, certainly smelled the gas and saw the rainbows in the water around their perimeters. They believed the oil had medicinal properties.

These "seeps," as they were called, were the first clues leading to the eventual knowledge that most of Louisiana, a good deal of Texas, and much of the Gulf of Mexico float on an unimaginably huge hydrocarbon reservoir. If you look at a map of the continental United States today, its orientation pretty much describes the general lay of the land as it has been since the Mesozoic Era some two hundred million years ago. Everything flowed down, or south, then as today.

Except then, much of what is now the southern United States was covered by an ancient sea that existed 150 million years ago, give or take 10 million years on either side. The sea eventually dried out, leaving salt layers miles thick in some places, what petroleum geologists reverentially refer to as the Mother Salt Bed. Over time, muck covered the beds up. Vegetation thrived, died, settled, over and over again during eons. More recently (100 million years ago or so), dinosaurs, prehistoric fish, and crocodilians inhabited this wet, steaming environment, far stickier than South Louisiana is today. Interestingly, some of the region's present fauna—alligators, bowfin, paddlefish, and garfish—to name a few creatures—are their descendants and have not changed appreciably since back then.

Sandstone and shale layers accumulated over this organic miasma, squeezing it, crushing it, twisting it, changing it to oil and gas over tens of millions of years. Some geologists say, with a hint of humor, that gas comes from plants and oil from dinosaurs. All the accumulating weight on the underlying foundations of the Mississippi River valley had two profound effects. One, its pressure sculpted a wide, deep trough called a geosyncline, an enormous compaction of part of the continent's spine. Two, sporadic ruptures, not unlike those of a herniated disc, burst upward. The pulp in this case was salt, great shafts of it miles in diameter, working its way upward, bending and breaking through sandstone, saturated with water, oil, and gas, like a mushroom bludgeoning its way through concrete. Hydrocarbons floated to the top of the bent sedimentary layers, leaving the heavier water below. But in some places, the salt heaved through the earth's surface pushing the flat land upward and allowing oil and gas to leak out. Five of these salt domes break the Louisiana marsh. Avery Island is one, the tip of a mountain of underground salt higher than Mount Everest. Hundreds of domes never made it to the surface, but their upthrusting still bent and broke the sedimentary rock, permitting hydrocarbons to rise upward, where they were more easily discovered.

Aside from its use as caulking and its supposed healing properties, oil was something of a mystery. If you happened to be a salt miner, it was definitely a bad thing; it could ruin a good day's work if it got into your take. But oil did burn. Salt miners in Kentucky along the Cumberland River found that out in 1829 when they were working a mine and oil gushed out of the shaft and oozed down the river. The miners did not know what to do with it, so they set it aflame and the whole river lit up like a serpent in agony.

Illumination was thus the initial main use for oil after the country's first well began producing in Oil Creek, Pennsylvania in 1859 (later renamed Titusville). Just in time, for whales were becoming restricted to the far reaches of the western Pacific. The next wells were drilled in California. The idea of producing oil in Louisiana didn't register, despite all the seeps in the marsh and oil sheens off the coast. It was a problem of accessibility. How do you dig a well in water and muck?

They didn't have that problem over in Texas. Spindletop was the country's first gusher, drilled in 1900 by an Austrian wildcatter named Anthony F. Lubich, who later changed his name to Lucas. An engineer, he spent years wandering the country until he ended up at Avery Island to apply for an advertised position in the salt mine there. Someone was needed to design shafts and maintain pumps to keep the salt dry.

Lucas's travels had taken him to California. He learned there what oil was. He may have seen the Santa Fe Springs oil field in the orange groves in southeast Los Angeles, the derricks going up so fast that they looked like they would soon outnumber the trees. He saw the oil around the base of Avery Island and smelled the gas in the air. But drilling there? Impossible. He had heard of a salt dome not far away, in Beaumont, Texas, in dry prairie country. On a hunch, he leased 27,000 acres around the dome, which he named Spindletop, and began drilling in late 1900, not with a rotary bit—that technology had not come into being yet—but by driving a shaft into the ground as a pile driver does. It was slow and expensive work. If Lucas had not been able to persuade J. M. Guffy, a partner of Andrew Mellon, to bail him out, he would have had to stop drilling.

Then, in one day in January 1901, Texas changed forever. The well blew up, shooting pipe five hundred feet in the air, followed by great belches of gas, followed by a geyser of oil—the country's first gusher. The Spindletop gusher shot oil two hundred feet into the air for the next ten days, polluting the land for miles around. After the well was brought under control, its oil was stored in open pits around the well. At one time, three million barrels-worth formed a lake with such perfect reflectivity that on a cloudless day, the blue sky seemed to fall right into it. Countless ducks and geese, mistaking the surface for water, skidded down to their doom. If the environmental movement had been in force then, Lucas would have been driven into bankruptcy for all the lawsuits thrown at him.

Jennings, Louisiana, in the prairie but not far from Avery Island, was the next gusher. Photographs of the Jennings oil field in 1904 show a forest of derricks. Their clutter represents the risky nature of drilling in those days. It was long before seismic shooting revealed the lay of the land underground. Drilling was guesswork. A rig could miss an oil trap by a couple of feet and come up dry. Stabbing the sediments in a frenzy was the only approach. There were a lot of misses but some substantial hits. The Jennings field is still producing. The Caddo oil field in north Louisiana was another discovery, a huge deposit discovered, full of gas as well as oil. No one knew what to do with the gas, so it was burned off, 27 billion cubic feet of it between 1908 and 1913.

The marsh remained elusive to drilling technology. Then in the mid-1920s, a Venezuelan ship captain named Louis Giliasso developed a submersible drilling rig. It consisted of a barge with a hole through, from deck to bottom, over which was erected a drilling derrick. The whole affair could be sunk in shallow water. The barge at rest on the floor provided a stable platform for drilling. The Texas Company—now Texaco—grabbed the man and the design and patented it. The challenge now remained to get this gear inside the wasteland. Early oil entrepreneurs went at it with enormous enthusiasm, changing the face of South Louisiana to this day. They lay boards across sections of *flotant*, or floating marsh, which dominates the interior wetlands. They engineered marsh buggies to churn over the marsh with great round iron wheels and later, with treads. But most of all, they dredged channels through it. The marsh put up little resistance. It was surprisingly easy and quick work, one canal for every drill rig, dug wherever needed, spoil banks hurled up in all directions. No one cared about blocking the marsh's sheet flow; no cared about ponding—the accumulation of trapped water inside the marsh that killed acres of vegetation. This land was unfit for human habitation and there were absolutely no laws against the early oil industry's unpremeditated redesign of the marsh.

There was the financial bonanza to consider. Despite the huge tracts of marsh controlled by individuals like Wisner, and companies like Sun Oil and Cities Service and Fina—plus scores of smaller outfits with such down home names as Pelican Oil and Pipeline, Prairie Mamou, and Spring Hill—the state still owned most of the marsh as a result of the Swamp Lands Act.[1]

By the early 1970s, Louisiana was producing over 800 million barrels of oil each year and around eight trillion cubic feet of gas.

---

[1] The state also controls most of the drilling in waters up to three miles offshore, considered Louisiana territorial waters. As wetlands, many of them privately owned, are encroached upon by the Gulf each year, the new bottomland reverts to the state, a source of considerable unhappiness among landowners. As the encroachment continues, the seaward boundary, however, remains fixed, providing the state with an increasing amount of bottomland. This beneficial arrangement was maneuvered by Senator Russell Long in an agreement with the federal government in 1954, known as the Tidelands Settlement, the conclusion of a three-decade-long lawsuit. The ocean bottom beyond the three-mile limit belongs to the federal government, referred to as the Outer Continental Shelf, or OCS lands. Royalties from oil and gas drilling in OCS waters go directly to the federal government. There arose the question, however, of royalties from "oil on the line," as it came to be known of fields that straddled the state/federal boundary. How would they be divvied up between the producing states and the federal government? By the time the question was addressed to the federal government by the producing states, over $6 billion had been paid out in leases and royalties. The government offered to return to Louisiana something over 8 percent of the money. Louisiana protested, saying it wanted close to 17 percent. The feds came back with a 3 percent offer. In a zigzag of challenges and numbers, offshore-oil-producing states contended that, actually, 50 percent would be an equitable amount, arguing that coastlines were being damaged by offshore oil and gas development. Predictably, the matter wound up in the courts which granted producing states—Alaska, California, Louisiana, and Texas—27 percent of the escrow amount and 27 percent of future leases, bonuses, and royalties. Louisiana feels burned.

Over 50 percent of the state's revenues came from oil and gas leases, taxes, and royalties. The Louisiana Oil and Well Distribution Map is peppered with tiny dots—green for oil and red for gas. Each dot represents one of the quarter-million-plus wells scattered through every parish in the state. The Monroe gas field up in the northeast corner is a solid blob of red.

The field day is drawing to a close, at least on land. All the wild-catters have gone to Africa and Latin America. There are no more "new elephants" in Louisiana, the term given to major discoveries. The last one was in the mid-1970s, a giant arc of underground gas with the magical name, Tuscaloosa Trend, that sweeps from east Texas through central Louisiana and into Mississippi. Now, Louisiana derives less than 30 percent of its revenues from oil and gas. And the future does not look good from onshore and bottomland wells. Present yearly production—150 million barrels of oil and 1.5 trillion cubic feet of gas—is predicted to dwindle to a forty million barrels of oil and a half-trillion cubic feet of gas in 2030.

The future of oil and gas is offshore, far out in the Gulf of Mexico, beyond Louisiana's territorial waters and beyond any direct benefit to the state through severance taxes and royalties. In 1998, OCS production (Outer Continental Shelf) off the Louisiana coast amounted to 400 million barrels of oil and four trillion cubic feet of gas, dwarfing that of any other state, even Texas. Forty thousand wells have been drilled out there in water approaching a depth of five thousand feet, their production coordinated by almost four thousand platforms—technology-rich behemoths standing in the Gulf on massive concrete legs. To a considerable extent, the federal government depends on the revenue from these operations to support itself. That Louisiana receives nothing in return as its wetlands die is a growing source of irritation.

If you wanted to study staging bases for future space coloniza-tion, Port Fourchon would probably be the nearest you could come on earth. This is the mother base of the oil and gas world out in the Gulf. On the coast and only fifty miles due south of New

Orleans International Airport, it seems nevertheless disconnected from the rest of Louisiana, though the man who runs it, Ted Falgout, is a Cajun to the core, and some of the people who built this perch at the edge of the Gulf of Mexico grew up in towns along Bayou LaFourche. Port Fourchon is not a pretty place. I don't recollect seeing one tree growing within its flat confines of dredged and reclaimed marsh. Dusty cars and pick-up trucks from Alabama, Oklahoma, Colorado, Tennessee, even a couple from New Jersey, clog its parking lots. They have been driven here through long lonely nights, by the 13,000 or so oil workers who jump from car to boat for the haul out to a platform for the next week or three weeks, or whatever their shift calls for. And then back to Oklahoma or wherever. This is not a Louisiana place. There's a bar but no dance hall, no families or crawfish boils, no place to hang a porch swing. This is a transitional station—a place you go to get somewhere else.

Without Port Fourchon, in all its rawness, like a new wound in the freshly dredged marsh, the source of 35 percent of this country's energy would literally founder. It is from here that most of the platforms out in the Gulf are supplied around the clock, in a growling of helicopters, crew boats, barges, and oceangoing tugs. Cruise the canals of the port—which is not a particularly good idea, particularly in a twelve-foot aluminum skiff, as I did—and you feel like a child crossing an interstate highway at the beginning of rush hour. The *Mr. Jack* and the *Miss Charlotte*, sturdy bulldog tugs, loom above, bearing cargo for the platforms—bright orange tanks of drilling mud, other tanks big and small, cranes, chains, pipes, floating pipes. The *Ram Charger* and the *Canyon Runner* lunge from the other direction up Belle Passe from the Gulf. The combined quad of herculean power makes a water spider out of anyone in their path. The *Mr. Joe* and the *Fast Cajun* rev their engines in a deep grumble from a bulkhead, surges of muddy water boiling from their sterns. Incredibly, a couple of porpoises undulate their way up the canal among all this hardware, the glistening curvature of their backs the only softness against the shuddering steel and sharp angles.

Warehouses with names like Tesoro, Halliburton, and Edison

Chouest mark the supply bases of this interpodal transport system. Ted Falgout is mighty proud of it all. He's seen it grow over the past couple of decades from a dumpy little shrimping port, with a deli inside a tilting mobile home, to this. Maybe that is why he is always smiling like a surprised kid to whom something unexpectedly good has just happened. A couple of old trawlers are still around, their gunwale planking gouged and pummeled, their nets drying soft and sensuous. They appear as ancient as *dhows* against their new backdrop of steel and cable.

"We move six thousand people a week by helicopter. We service two hundred fifty vessels in this port daily. We've got a billion dollars' worth of boats here. This is a hidden city," Falgout gloats. The latest addition is a facility called C-Port, a gigantic covered warehouse into which huge supply boats can slip, unload, reload—with the help of some robotic-appearing Gantry cranes—and head back to the Gulf in a ten-hour turnaround time. Falgout calls it the "Wal-Mart of deep sea services."

There is only one way to get to Port Fourchon: Route 1. It's a snaky little road in most places that shadows the curves of Bayou Lafourche, as it winds its way down through Cut-Off, Golden Meadow, Leeville, and through the marsh to Fourchon. One thousand six hundred tractor-trailers wind down this road every day, rippling its asphalt into wavelets of hard riding. They carry all sorts of oil field paraphernalia—heavy pipes, drilling mud, huge engines, and the like. Over the years, they have squashed Route 1 down into the marsh. Of course, the marsh itself has subsided.

Then there are the old oilfields beside the road—the Leeville Field, the most pervasive—where the marsh has been so cut and diced by canal dredging that it looks like a ratty checkerboard. The canals dredged through this section of marsh over the years have pretty much done away with the marsh. Spoil banks in parallel lines and earthen barriers at the end of keyhole canals are the only landmarks. Waters get all bottled up in this maze, can't flow one way or the other, which only further erodes the marsh after the vegetation drowns. "Us Cajuns didn't ask for all this development," Falgout says, referring to how the land was carved up and left in furrows. "World War II came along and they needed oil and gas. That was in the national interest," he says, wearing that smile, and one begins to wonder how he can keep it going as the marsh around him crumbles and disappears.

As Route 1 approaches Port Fourchon, it seems like you are about to plane right into the marsh's skeleton, or into the Gulf. There's not much on either side of you, except water that laps at the road's shoulders. You feel that you are as much a part of the marsh as the egrets out the vehicle window or the blue-winged teal jumping up from a ditch beside the road. I have never driven an eighteen-wheeler down Route 1, but if I ever did, I think I would be scared. I would think that the high cab would plunge right into the water, especially going around a curve with the trailer carrying three or four tons of steel and exerting not inconsiderable force on the cab's momentum and direction. It would especially occur to me that this asphalt leading to this port on the edge of the world is nothing more than a frayed thread that the slightest quirk of nature could snap in a second.

Port Fourchon itself is not much better off, sticking out into the Gulf as it does, the Gulf rising, the marsh around it messed up and sinking. Falgout takes me out to the beach beyond the port, the real edge of the Gulf. A hundred feet offshore, a line of a dozen or so stone-filled old barges hunker in an arc around the headland where we are standing, planted there as guardians against eroding storm surges. Falgout says he had them put out there after Hurricane Andrew blew through. The coast has receded since then, closing in on the port. "It just killed me that Dan Rather got on a beach on Long Island and said that they are going to lose forty-five miles of sand due to rising sea levels, as if that is the worst of it. Right here we are going to lose forty-five *square miles* of marsh," he says, smiling. In a more recent effort to fight the encroachment, Falgout has made up a batch of concrete ajax blocks which line the little dunes down behind the beach. They are of the same design as kids' jacks, but in their gray drabness and waist height, they look like land mines. Falgout said he read somewhere that they are using them to protect coasts in Europe, so he thought he would give them a try here.

When you first meet him, Falgout appears ready to boast only about his port's attributes. It takes him a while to get going on its vulnerabilities. But then he warms to the subject and just takes off. "If the marsh keeps sinking and the beaches keep deteriorating and we get one big hurricane here, Port Fourchon is gone. It'll make the OPEC crisis in 1974 look like nothing. Then, how are you going to assure the guy up in New York that he can keep his furnace on?" His smile broadens like a bad boy who has made a good point. Then he picks up again. "So, if all the pipelines bust down here, I guess we could still keep going with all our fisheries." Then a thoughtful scowl darkens his eyes for a second. "Of course that might not work because all the fish would taste like oil. Anyway, we couldn't transport them out of here because Route 1 would be knocked out." He smiles again, having made another pointed but foreboding point.

Oil Companies are supposed to clean up their messes. Some scientists say that the demise of Louisiana's wetlands is 90 percent the fault of oil-company canal dredging. Most scientists say that this is an exaggeration. They say that the combination of the Mississippi's levees, the rise in sea level, coastal erosion, and saltwater intrusion are all contributing factors. The fact remains, however, that the companies are legally obligated to restore any area they drill on or transport oil and gas through. The state Department of Conservation, a branch of the Department of Natural Resources, is supposed to require them to do so. The office has even set aside a fund so the state can pay for wetlands restoration when an oil company, a wildcatter gone bankrupt, say—is unable to.

Restoration is one thing in fact, another in practice, and highly subject to interpretation. John Englhardt, who carries the inimitable title, Abandonment Business Manager of the Chevron Environmental Management Company—a Chevron subsidiary—agrees with the regulations. "We have an obligation to return the land to its original state." "Original state," to his thinking, does not go so far, however, as to refill old canals or get rid of spoil banks. It means cleaning up oil field equipment—storage tanks, pipes, valves, and the like. It means doing what he calls "P and A"s, on depleted oil wells—plugging and abandoning them. It means filling in production pits, and dredging contaminated canals.

Chevron Environmental Management was formed as a Chevron subsidiary in 1998 to devote itself wholly to this kind of work around the country. Englhardt, a cheery man who exudes pride over Chevron's effort to clean up after itself, spends a lot of time traveling from one Chevron facility to another from his headquarters in New Orleans. His biggest present challenge is to oversee the "P and A"s of 1,700 wells in Bakersfield, California, a project that was getting under way in mid-2000.

To Englhardt, "original state" also means returning the land to its intended use, "whether," he tells me, "it's for farming or housing." Intended use for wetlands? Wetlands have never had an intended use. They are just there. But he argues that oil companies would, in fact, be doing a disservice to the marsh environment by taking the marsh back to its state prior to the oil industry. Many of the canals were dredged a half century or more ago. A wide array of animals—deer, coyotes, nutria, mink, otter, fox, and raccoon—and fish—redfish, bass, and speckled trout—have adapted either to spoil banks or canals, making both of them artificial but well-established econiches. "You get some of the best fishing in the state in those canals," says Englhardt. To destroy them and the spoil banks would result in enormous ecological havoc.

This perspective notwithstanding, oil companies are generally more attentive to their disruption of the environment than they used to be. But that does not mean much, considering the magnitude of their earth moving. Around 1,000 acres of wetlands still disappear each year into the maws of dredges gouging out pipeline and service canals. Gulf Liquids New River Project, for example, a Houston-based company, recently completed an eighty-mile pipeline from an Exxon/Mobil refinery in Chalmette—downriver of New Orleans—underneath Lake Pontchartrain, and through forty miles of cypress and tupelo gum swamp to a fractionation facility—which separates natural gas into components—in Geismar, upriver of the city.

The pipeline has impacted over 400 acres of wetlands, according to the Lake Pontchartrain Basin Foundation, a grassroots environmental organization with a broad-based membership and considerable clout. Initial designs called for it to be laid through virgin swamp. "This is land that maybe no one has ever set foot on," says Neil Armingeon, the Foundation's environmental director. "It's not that oil people are so bad. They just have what I call a Texas mentality." In Armingeon's view (he is from North Carolina, where he says people fight to preserve two acres of wetlands—quite different from Louisiana where two acres is a blip), this mentality is particularly poignant in the case of the Gulf Liquids New River Project's pipeline.

A pipeline corridor maintained by Entergy and utilized by Equilon—a Shell and Texaco enterprise—has long passed through the area. According to Jim Taylor, manager and owner of the Gulf Liquid's project, Gulf Liquid's apparent hard-line attitude against the environment has nothing to do with a "Texas mentality." Initially, Equilon did not want Gulf Liquids to share the right of way. "There was a time we thought we would have to clear trees from the marsh because of Equilon's resistance," Taylor explained. Why the oil and gas industry cannot automatically share right-of-ways is beyond Armingeon, as it should be. Only after considerable persuasion by the Foundation was a sharing agreed to, which saved almost twenty acres of old-growth cypress. "One of the joys of being in Louisiana," Armingeon says, flicking his ponytail, "is that this state is not really part of the first world, the western world in the twenty-first century. At the same time, one of the horrors of Louisiana is that this state *is* in the first world."

From Six Mile Lake, adrift with islands of wind-blown water hyacinth and the sweet smell of willows in spring blossom, you enter Little Bayou Joe lined with glistening white spider lilies in full bloom. The day is brilliant, a dry north wind and some good rains up in Ohio driving some water down the Atchafalaya River. It gives hope to people like Peanut (whom you will meet more extensively in chapter 6) whose main occupation with the coming of spring is crawfishing in the Atchafalaya, the largest swamp in America at 1.5 million acres. Then you hang a left into Big Bayou Joe, wide and muddy, its edges obscured by alligator grass. Then American Pass comes in from the left and you veer into that. There are a couple of camps there. One has a bright red playground set next to it that looks out of place out here among the huge moss-draped cypress just busting out in new foliage. Then Peanut, whose real name is Sidney Michel, turns left into Bayou Boutte. Half a mile or so, another left brings you into a canal with no name.

Everything suddenly changes, as it did for Alice falling down the White Rabbit's hole. Except here it's sideways rather than down, and it's instant, like passing through the wall of a bubble, from paradise to a demihell. Paradise is outside and the things of paradise can enter and leave. But inside is a bad place. Oil smell thickens the air, greasy and clinging. The spoil bank on one side has a pipe up on pilings running its length and if you look carefully behind the emerging leaves, you can see all sorts of rusting pipes, hunks of iron, splotched and dented fuel tanks. Red-winged blackbirds cackle their alarms from the willows and a couple of egrets in the shallows break their poses to glance at the boat, regain their composure, then fly off. The names EXXON and then HILLCORP ENERGY CORP. appear on pipes and on a well protruding from the water.

Peanut says people around here have always called this the Exxon-Duck Lake Oil Field. Been here ever since he can remember, he says, which means that it is at least thirty years old. Actually, the field was opened in January 1949. Hillcorp, a Houston-based firm, bought it from Exxon (now Exxon-Mobil) in 1994.

Ahead of us, big squat holding tanks are bunched up on the spoil bank with a dock in front of them and a barge with a cabin fastened to its deck. No one's around. A little further on, more tanks are up on the spoil bank, these towering and narrow and tethered to each other by a maze of horizontal and vertical piping. Every few seconds one of the tanks wheezes a pff-pfff-pffff sound, louder than any of the birdsong nearby or the gra-umpf of bull-frogs. It sounds as if it were having pulmonary trouble. Corrosion has eaten the tanks' walls and some white stuff oozes out of the cracks. It looks like there is going to be a meltdown any second. Peanut says it's just a rotted coating of insulation that makes them look that way.

What's on the spoil bank doesn't look healthy but not as bad as what's behind the tanks on the other side of the bank: a graveyard of dead cypress trunks jutting into the air like drunk telephone poles. They're sitting in water, maybe twenty acres of them—a big expanse of emptiness in America's biggest swamp. In one place, there's a cluster of naked trunks and a Louisiana heron sits on a branch. Surprisingly, it looks just fine.

Peanut and I had been out collecting crawfish from his traps nearby. Low water made for a measly harvest, sixty pounds or so from a hundred traps that Peanut had sunk along a trail that wound among huge ancient cypress that had escaped the lumbermen a century earlier. It was spiritual back there, floating tiny among these trees and the spring vegetation. We were the only two people on earth, it seemed, and the earth was not much aware of whether we ever emerged or just got lost. A horned owl hidden in a cypress's sphagnum moss kept hooting at us. Peanut said it was marking where he raised and lowered his traps. In a few hours, it would swoop down and lift one of them by the string, talon over talon, until the top of the trap was above the water's surface. Then the crawfish would come up for air, all crowded together in a little space, and the owl would reach in between the chicken wire and grab them. But this owl was not going to get away with that trick. Peanut had a way to outsmart the owl. As we wound through the swamp, he left bits of baitfish on tree branches or on logs. The owl would grab the bait and forget where the traps were. Finally, we saw the owl when we were just about under the cypress. It flapped out soft on silent wings like a huge brown moth and the trees beyond swallowed it up. Peanut chuckled. He likes owls and raccoons and otter and other smart animals.

I had seen the oil field briefly during the winter when the swamp was dead and sad-looking. Now I wanted to see it while everything was bursting with life. When Peanut was finished emptying his traps,

I asked him if we could ride over there. Drillers used to be able to dump all the junk that comes up in the drilling process—saline solutions, radioactive mud, and a variety of toxins—right next to the well as long as it was enclosed by levees. What I am looking at is a reservoir of waste, enough to make an environmental activist go apoplectic. Production pits like this are all over the state, wherever older gas and oil fields were developed.

It could be much worse. I am, in fact, impressed that it is not. "It's jus' an old pit," Peanut comments. "They used to dump all the stuff back there. That's jus' where they put it." Whatever toxins are still back there could have made this corner of the swamp a sickeningly dead mud heap. Some of them are like that. Nothing grows. But it's not that way here. Vegetation juts up from the water beside the dead tree trunks. I can see egrets back there wading the edges of the pit. And everything beyond the edges, miles and miles of everything, looks just fine, as if this was just a healing scab on nature's flank.

We cruise further on up the cut and come to a ramshackle building, right across from a huge rusting oil tank perched up on a bank, vegetation trying to hide it. The tank noses out of the sand like some strange emerging thing with dark stains on it. The building is of gray corrugated metal, rusted roof, windows broken, door ajar. Junk pipe surrounds the place, enormous rusting valves and couplings. Some of the pipes descend into the ground. Gauges fitted onto them show high pressure within. Someone is supposed to be living here; somehow I feel that. Maybe it's the dirty broom outside on the little dirty deck. Maybe it's the way the walkway angles up to it from a dock that has half sunk. A crud-stained aluminum boat with an outboard is drawn up on an oil-stained bank outside, looking so unused that it might be ownerless by now. I don't want to go into that building. I don't want to think what a person who lives in this place would look like.

Peanut's pretty quiet here. We cruise around a bit more, silent, me listening to the wheezing of the pressure-release valves on the old rotten-looking tanks and looking at the dead trees, and wondering why Peanut seems surprised that I am interested in this place. I think I know why. The Atchafalaya Swamp is over a million acres of mystery, a cathedral unlike any other natural place on earth, a place where most people just naturally go quiet. Being quiet out here is just what you feel like being. Peanut spends days and weeks out in the swamp by himself during crawfishing season. Given a million-plus acres, this devastated twenty-acre lot is really nothing more than a forgivable pimple on otherwise perfect skin. Of course, there are other pimples, hundreds of small oil fields and gas fields, but those aren't visible from here. Nor are the toxins in the ground. And then there does seem to be normal, if sparse, life emerging from the toxins. Peanut, like so many people in this part of the country, tends to look at the abundance all around him, just outside of this particular bubble of hell. He tends to ignore the actual presence of hell.

Some people here are losing patience. Michael St. Martin, who calls himself "just a Cajun lawyer," is such a person. He grew up in the oil patch and marsh just south of Houma. His grandfather built oil rigs; his father ran oil service companies; St. Martin drove tanker trucks to work his way through law school. Now fifty-six, he makes impressive money as a personal injury lawyer representing oil workers hurt on the job. Behind this professional heritage stands the marsh, which in one way or another supports just about everyone in South Louisiana. He hunts alligator much of each September and when duck season begins, the draw to get out on the water in a cane blind with his dog is just about irresistible.

Like most people here, St. Martin is angry about the disappearance of the wetlands. Unlike most, he has decided to do something about it on his own. Armed with his legal knowledge, considerable money, and a willingness to thumb his nose at the oil industry, he is flogging them into court, something that no one, surprisingly, has ever before had the gall to do. His name invariably comes up in conversations about the state of the marsh, as in "St. Martin has the oil companies on the run," or "that St. Martin fellow has a real game going; bet he plans on suing every oil company in the state." The comments are not all favorable. While there's relief that at last someone is doing something about the canals and spoil banks and erosion, not a little embarrassment is evident, too, that a local boy is biting the hand that has fed the local people for generations.

St. Martin's means of revenge is the marsh. In 1992, he bought up 7,000 acres of surface rights for $245 per acre just south of his home, which he calls Mandalay Plantation, a name taken from an antebellum sugarcane plantation that once operated nearby. Mandalay is now a suburb of Houma on Bayou Black. A quarter mile up the bayou is another suburb with the oddly contrasting name of Waterproof. Some people think St. Martin is buying up the marsh only so he can sue the oil companies that have damaged it. That's not entirely true. He bought the acreage in agreement with the Nature Conservancy; the Conservancy would take over 5,600 acres with the financial help of Dow Chemical and deed it to the federal government. The area is now the Mandalay National Wildlife Refuge. Using the remaining 1,400 acres as evidence, St. Martin went to war against the oil companies.

I meet him in his New Orleans apartment and office complex so he can tell me about it. His place is in a recently renovated warehouse on St. Peters where, not entirely happily, he says, he perches two or three days of the week doing business before getting back to his beloved marsh. He lives on the sixth floor—the top floor—and as I ascend to his aerie in the glass-walled elevator, I gaze across the atrium of the gutted and rebuilt warehouse, down into the greenery below. I smell serious money. He comes to the door, tall and lean, with a mop of gray hair atop a chiseled face. He's wearing a faded red tennis shirt, new white pants, bright red socks, and deck shoes. The first thing I see in the apartment is a halfsize nude of translucent lavender glass. She is sitting on the edge of a table starring at me as I enter the living room. She's beautiful. Her expression is of fetching innocence. Her breasts are perfect. Her thighs are perfect. Disconcertingly, that's all there is to her.

A menagerie of animal skins covers the floor of the immense living room with soaring ceilings. A zebra has its rear toward a mountain goat that is hiding under a chair. Antelope of various species have gathered around the gleam of a polished dining table and a Kodiak bear faces a window from which, if it were more than a skin and gigantic head, it could spy the river and an edge of the French Quarter. A huge wolf has draped its luxurious pelt over a couch and I sink into its softness. St. Martin says he shot the animals—in Africa, Alaska, the Rockies—all except the bear. A friend gave that to him, he says.

The legal issue is 357 acres of damaged marsh, St. Martin tells me, jumping up to fetch a copy of Judge Marcel Livaudais, Jr.'s ruling. Back in the 1950s, Superior Oil Company, the predecessor of Mobil Oil, leased the mineral rights to St. Martin's land, later selling them to Phillips Petroleum. Land ownership in Louisiana is complex. One can own a piece of property in its entirety. Or one can own just the surface. Or one can own just the mineral rights. Or one can own one and lease the other. While first Superior and then Mobil and then Phillips leased the mineral rights, St. Martin and his wife purchased what they thought were just surface rights from a company called Southdown Exploration, a former sugarcane concern reinvented as a land and gas and oil company. Before St. Martin came along, Southdown had also granted what are called "canal servitudes" to Superior, meaning that the oil company had purchased permission to dredge a series of canals through Southdown's property in order to drill for oil or gas. These access canals entered the property from the Intracoastal Waterway.

St. Martin noticed that every time a barge traveling the Intracoastal crossed the openings to the canals, wash surged up them and into the marsh—now his marsh—and surged out again, tearing the marsh to pieces over the years. He wrote several letters to Mobil requesting that it plug some of the canals so that boat traffic wash would be prevented from flooding into the marsh. He becomes agitated as he tells me this, hands moving in a flurry, legs crossing and recrossing. I can see him in court arguing injury cases before a jury, though this case never went before a jury. He is, I imagine, very persuasive.

"Statutorily and contractually, oil companies are required to restore land," he booms out, his menagerie now his jury. "If an oil company damages a sugarcane field, they'll repair it. But the marsh—it's always been taken for granted."

Mobil didn't take much of an interest in the matter, St. Martin says, except to remove some oil drums and pilings, and install temporary canal plugs that washed away in less than a year. His face darkens as he tells the story. "I said to their lawyer: 'Hey, sport, what are you going to do about that problem?' He told me that they weren't going to do anything.

"They treated me like I was someone lost from an insane asylum," an article in the *Times-Picayune* quoted St. Martin. "They told me 'No one has covered up or restored oilfield locations or pipeline canals in Louisiana.'

"They've taken $135 million of hydrocarbons out of that marsh through those canals. It would cost them $2 million to restore the damage. . . ." St. Martin grimaces through clenched teeth, letting the horror sink in. "They wouldn't even offer to settle."

That snub may have changed South Louisiana forever. St. Martin and his wife sued both Mobil and Phillips, the present lessees, for the restoration of the most damaged part of their marsh. It was a novel lawsuit. People might have thought of suing before, but had never done so out of a perverse obeisance to the oil industry and out of realization that when the oil and gas people carved up the marsh half a century ago, no one had realized the consequences. So no one was really to blame. It was like General Electric dumping PCBs into the Hudson River during the same era—the 1950s—when the idea that there could be far-reaching consequences to the environment had not yet dawned and when there was no one watching, anyway. Rachel Carson would not alarm the nation with her *Silent Spring* until 1962. PCBs and other strange chemical compounds were too new for anyone to suspect their role in causing disease and birth defects. In the same vein, there were too many wetlands for anyone to think that a canal here and a canal there could add up to something approaching ecological destruction.

Along with the lawsuit, St. Martin presented a restoration plan for his marsh. He asked that Mobil and Phillips see to it that permanent levees be constructed where the canals met the Intracoastal Waterway. The levees would stop tug and barge wakes from washing into the marsh. He also wanted the defendants to plug unused canals and to remove wellheads—consisting of substantial pipe—from depleted wells. The bill would come to $39,000 per acre, or just under $14 million.

Judge Livaudais wrote in his eventual ruling that this kind of money was "grossly disproportionate to the value of the property." He noted that the St. Martins had paid only $245 an acre for it. He also ruled that the St. Martins could not sue for damages sustained before 1992, when they bought the property. The court said that it had determined that only twenty-four acres had deteriorated at the hands of Mobil and Phillips since 1992.

But was this damage entirely the fault of the oil companies? St. Martin called in wetlands experts who said that the canal dredging was 100 percent to blame for the marsh's condition. Mobil and Phillips called in wetlands experts who said that blame could be cast to a litany of ills—hurricanes, saltwater intrusion, nutria, as well as canal dredging.

In the end, the judge declared a draw of sorts. Sixty percent of the damage, he ruled, was due to the oil companies' negligence, 40 percent to natural processes. The award was disappointing—$10,000 per acre, and only for twenty-four acres. Disappointing but a step up from $245 per acre. Mobil and Phillips have appealed. The St. Martins have also appealed. In the meantime, St. Martin walked away with $240,000, which he quickly spent plugging the canals himself and shoring up the banks of the Intracoastal so tugs' wakes wouldn't chew away at the marsh quite so quickly.

The case was only a foray into the unexplored. He has filed another lawsuit, this one against Koch Gateway, a pipeline company that St. Martin accuses of "ruining" another sixty acres of his marsh. Again, the decision to sue was based on a snub. Koch Gateway maintains a permanent right of way across a section of marsh for a pipeline canal. The pipe at the bottom of the canal carries 200 million cubic feet of gas per day for which it receives, says St. Martin, roughly over a half-million dollars per day. The canal is supposed to be forty-five feet wide but erosion has eaten away at its banks—St. Martin's marsh—to create a 110-foot-wide swath of open water. St. Martin wants to stop the continuing loss by plugging the canals. "It would have cost them $3,000 to $4,000 for the work and saved twenty to twenty-five acres of marsh," St. Martin tells me. Koch refused.

A month later, I am at the juncture of this canal and the Intracoastal. A tow has just passed—a tug pushing three barges weighed down to within a few feet of their hatch covers. The wake rolls in almost immediately and churns up the canal—a half-dozen two-foot waves. They smash against the canal's banks. Trees on the edge have toppled over into the water. The banks are raw and jagged from the constant erosion. Ricky Felio—a friend of St. Martin's who has taken me to see the damage—says that tows come through every five minutes or so. Three more are approaching.

We fly up the access canal in an airboat and Felio, a former oil-field worker and unabashed admirer of St. Martin, points out breaks in the spoil banks through which water from the marsh gushes like blood from a new wound. The tows' wakes have done that. The incessant rise and fall of the water has eaten away at the muddy banks, sucking water out of the marsh wherever it has bro-ken through. He runs the airboat over a spoil bank and we skitter toward a far corner of the marsh. The water becomes shallower and shallower until we are gliding across acres of glistening mud. We stop in the middle and look at the mess. Felio praises St. Martin's efforts to regain his land. "They're sayin' that this is a coon-ass lawyer with some brass balls," he extols.

St. Martin's rising anger at the obstinacy of the oil and pipeline is reflected in the damages he is claiming from Koch—$140,000 per acre for sixty-five acres. The amount reflects an ongoing rejig-gering of the value of wetlands, work begun decades ago by Eugene Odum, an ecologist at the University of Georgia. Odum exploded the myth that wetlands are wastelands. He put into the scientific literature what every boy with a minnow net knows—that tidal pools and estuaries are nurseries and crucial to the food web. They also sop up pollutants. Odum's work has now been eclipsed by ongoing research that has geometrically increased the value of wetlands. St. Martin based his valuations on the work of Robert Costanza, an economic ecologist at the University of Maryland, and his colleagues. It's a voyage of discovery, perhaps comparable—in an intellectual sense—to the thrill of threading your way down a bayou, marsh on either side, and not quite knowing what surprise awaits around the next curve. Costanza has assigned seventeen "ecosystem services" (he calls them) to wetlands. Nurseries are included, of course, along with food and raw material production. Minerals like oil and gas are *excluded* because they come from beneath the surface. Once ignored but now revealed are other values—the capacity of wetlands to regulate water flow, form soil, regulate climate, enhance tourism, and contribute to a sense of spirituality. The value of only four of the seventeen services have been calculated for Louisiana. The services that tidal marshes in other areas offer have been dissected in greater detail with the highest dollar figure coming in at around $9,000 per acre *per year.*

This amount does not take the future into account. While multiplying annual value by X number of years will render an enormous figure per acre, the future is unpredictable. In an economic accordion-like maneuver, Costanza has compressed "the stream of future services" into the present, coming up with a figure of $133,000 per acre of marsh. This amount, when combined with the value of gas and oil that might lie under the surface, gives some argument to St. Martin's $140,000-per-acre assessment.

Whatever the outcome of his lawsuit against Koch Gateway, South Louisianans are steadily aligning themselves behind St. Martin's tactics. Local school boards have shown the most excitement. In Louisiana, school boards own a lot of marsh, given to them early in the nineteenth century by the federal government as a means of school funding—rather paltry until oil and gas were discovered—in place of property taxes, which in Louisiana are negligible. The Terrebonne Parish School Board has hired St. Martin to represent it in suits against several oil companies that have cut canals across the marsh. St. Martin says that five thousand of the board's nine thousand acres have been destroyed. The case is scheduled to be heard by an elected judge in a state court. St. Martin is looking forward to the occasion. "It's going to be interesting to see if that judge will deny those school children money for their education or if he will allow the oil companies to get off without paying."[2]

Not all of the oil and gas industry is as stubborn as St. Martin's antagonists. Some companies have become innovative at cleaning up their messes. Of course, they still need a little nudging. Cathy Norman is a supreme nudger. It was partially due to her persistence—and legal threats—that Chevron formed its subsidiary, Chevron Environmental Management Company. Chevron leases seventy-five acres from the Wisner Donation, right next to Ted Falgout's domain at Port Fourchon. It's an elaborate production facility—a maze of canals edged by storage tanks, compressors, pumps, and what are known in the trade as heater-treaters, furnace-like tanks that separate natural gas into separate components—kerosene, benzene, etc. Most of the facility is devoted to receiving crude oil from Marchand Bay, the platform-studded section of the Gulf off Grand Isle, and sending it on its way via pipeline to the Chevron refinery in Pascagoula, Mississippi.

What happened at this facility is the beginning of a happy story that shows how well the oil industry could mesh with environmental concerns. When Chevron's lease came up for renewal in 1989, Cathy Norman, as manager of the Wisner Donation's land, insisted on a few changes, least among them an increase in rent. More important to consider were the contents of some old production pits, unlined reservoirs typically two hundred feet across where oil from wells, along with water that came out of the ground with it, used to be stored. Over time, the oil floated to the surface and was pumped off. The remaining water—called "produced water"—was ugly with such "aromatic" hydrocarbons as fluorene, naphthalene,

---

[2] At a pretrial hearing in late 2000, however, the court ruled against the Terrebonne Parish School Board, citing a clause in its contract with two pipeline companies stipulating that the companies would not be accountable for damage to the land.

and phenanthrene. Heavy metals like barium, nickel, vanadium, and zinc were also part of the mix, all suspected of causing a litany of ills from miscarriage to cancer. The concoction eventually settled to the bottom, accumulating mats of black ooze up to nine feet thick.

Fortunately, production pits are no longer in use. Separation devices—which separate oil from water—on oil platforms have replaced them. In 1993, the state declared pits a health risk and halted their use. Getting rid of their contents was not so easy. Before the prohibition, the state permitted the oil companies to merely hide the pits by pushing in the surrounding levees and capping the pits with a two-foot layer of clay. Companies jumped at the cheap, quick fix. If they didn't, the state promised to make it more difficult for them. It threatened to force companies to remove the contaminants, a hugely expensive job.

Hiding by capping was not good enough for Cathy. She predicted a disaster. She was especially worried about some pits a short distance from the Chevron production facility. They had at one time been back in the marsh. But with the incessant shoreline erosion each year, the pits had ended up right on the beach. The next storm could open them up, spewing tons of guck into the Gulf.

"We've got a big problem here," Cathy told Chevron, represented by John Englhardt, Chevron's Abandonment Business Manager. It was a problem that had to be solved before a new lease could be negotiated. The pit issue became part of the negotiations. She asked Chevron what it planned to do. Chevron knew that it wasn't just the lease renewal at stake. If any of that oil residue leaked into the Gulf, the resulting violation of the Clean Water Act could cost Chevron millions of dollars in fines. An agreement to dredge and fill the pits was quickly made.

But there were other pits to think about—twenty-one additional ones, all uncovered, which surrounded the production facilities. "During storms, they blasted oil across the marsh," Cathy said. Before any new lease was going to be signed, those pits had to be cleaned up, too. The matter might have become a sticking point if investigation of the pits' contents had not revealed something worse than anyone had imagined—high radioactivity.

Radioactive substances, uranium for example, occur naturally. Sometimes, they come up from the ground with oil. The pits around Chevron's facility were full of them, as was Dead End Canal, a quarter-mile-long service canal. Called NORMs, for Naturally Occurring Radioactive Materials, they included radium, polonium, and lead, consituents of the particular geologic formation underlying Marchand Bay. When brought up in a well pipe, two things happen to make them far more dangerous. First, they collect on the inside of a pipe like plaque on an artery. Drill pipe is a commodity in the oil patch, withdrawn after a well is depleted, taken to a pipe yard, cleaned up, and often resold. Cleaning is accomplished by "rattling pipe," as it is called—inserting a wire brush in the pipe and rattling loose the accumulations. What comes out is a soft, greenish flaky material. Workmen used it to fill holes in pipe yard roadways that heavy equipment had roughed up. It compacted nicely, creating a smooth lethal surface.

Second, NORMs concentrate in production pits, raising levels to the danger point. In the early 1990s, the industry found itself with a big problem. The state demands less than five picocuries per gram of radium in oil fields. Some of the pits measured in excess of 160 picocuries. Department of Environmental Quality inspectors, newly equipped with Geiger counters, discovered that pipe yards were buzzing with radioactivity. The same was true for Chevron's production pits. While the radium in pipe yards could be contained by changing cleaning methods, the pits presented a huge problem, one that was not going to disappear under a clay cover. The radioactivity had to be disposed of. That meant that each pit, plus Dead End Canal, had to be dredged clean and its bottom contents—a nine-foot-thick accumulation of toxins—trucked away to a disposal facility in Texas. Chevron officials calculated that there were 600,000 barrels of toxins to be carted. The price would be $300 per barrel for a total cost of $180 million.

Another alternative was to return the toxins to where they had come from in the first place—underground. It was a novel approach, an idea that brought Cathy Norman, the landlord's agent, and John Englhardt together in a collaborative relationship in which everyone—Chevron, the Wisner Donation, and the environment—came out a winner.

The question was where to put the well. Initially, Chevron wanted to drill offshore. The state provided a permit. Fortunately, wisdom prevailed with the thought that barging radioactive substances through the Gulf's chop would be foolhardy, given the ever present risk that a tow might break loose in a storm. At the least, it might send the wrong message in light of Chevron's public relations campaign as an environmentally friendly company.

Cathy's land was the next alternative. Chevron wondered if the well could be drilled on Wisner land. She was given $10,000 to research the possibilities, and hire geologists and NORM disposal experts. "What amazed me most was that this project required innovation," she told me. "You don't see much of that in this industry."

The result was the largest injection well in the world, five thousand feet deep into a sand bed sandwiched between impermeable shale layers. For six months, beginning in late 1997, a million barrels of the toxins were scraped from the bottom of the pits and the canal, mixed into a watery sludge, and forced down the well and through the sand lens. The amount is comparable to a football field eleven stories high. The project cost Chevron $60 million, a $120-million savings off transporting the toxins to Texas and no guarantee that they would arrive safely. And in a neat deal, the pits were filled in with clean material from a new slip that Ted Falgout was having dredged for nearby Port Fourchon. The port would have had to pay someone approximately one million dollars to carry away the spoil.

Cathy got everything she wanted and more for the Wisner Donation—a new lease on her terms, with $50,000 at signing, a $10,000 research fee, and, best of all, a $1.50 royalty fee for every barrel of slurry sent down the well, an easy $1.5-million increase in the donation's income. "Chevron even gave me a glass bowl," says Cathy happily. "We're friends now."

Chapter Five

The Prince of Shrimpers

George Barisich sits in his high seat in the cabin of the *FJG*, a fifty-two-foot shrimp trawler that looks like hundreds of others that fish the Gulf's inshore waters. She's glorious, as most trawlers are—high, sassy bow flaring back to gently curving gunwales that flow back to a broad stern. Viewed from the water or the banks of a bayou, the cut of such craft imparts to them a startling grace, the same as often observed in some stocky people who manage to transform a stumbling misstep into a pirouette. Vulnerability is part of their beauty. Whether trawl outriggers are lowered, supporting the great nets that grate along the Gulf's bottom, or raised for passage home, the massiveness of the rig dominates the boat's curves, suggesting the inevitability of a disaster that rarely occurs.

On board, raw strength of steel, cable, and winches crammed onto the wide deck obscures the *FJG*'s grace. The focus here is clear: fish smell hangs heavy; oyster shells and dead crabs crowd the scuppers, flung out of dripping trawls and ignored; a mobile of insatiable gulls hang above us, shifting their balance with the boat's heaves and groans through the swells. This boat is a workhorse through and through; she is only accidentally beautiful.

George's hands rest lightly on the wheel spokes while his eyes flit over the Gulf's waters ahead. Clouds build on the horizon to the south. It's nice out here, even in the middle of July, a cool wind blowing off the swells, diluting the heat, the water tinged in green, the heavy rigs astern bumping along the bottom which fills with whatever is down there and makes the boat grumble as though she were trying to talk.

George is the main shrimper in his family, trawling off the Mississippi coast a week at a time out of his home in Violet, Louisiana, a river town below New Orleans. He shares the *FJG* with his brother, Joseph, and his sister, Frances, hence the boat's name. I am on board for a few days to witness, and take pleasure in, a traditional business that is dying.

My eyes keep going back to the wheel as I stand next to George, trying to figure out how he knows where the shrimp are. The wheel is elegant, of teak layered in curved sections joined by brass screws, spokes angling out symmetrically, each dully glowing with sweat and oil. Its apparent age suggests the passage of great things—battles, storms, and voyages to uncertain destinations. Actually, the boat that it directs has experienced nothing more than what any other workaday trawler would. But this wheel and the man whose hands gently caress its spokes do have stories to tell that keep fresh the thirty-five years he has spent on the water, ever since he was eight.

The stories have become shorter and angrier over the years and punctuated by worry. The price of shrimp has not risen in a decade. Everything else has gone up—fuel, ice, insurance, regulations. And the boat is getting on, forty-two years old, a year younger than George, beautifully crafted of cypress that you can't find anymore. If anything happened to her, George could never afford the half-million dollars that a new steel boat would cost. Nets are $850 apiece, easily ripped to shreds by junk on the bottom. Demand is way off; some 80 percent of the shrimp consumed in this country is imported, mostly raised in artificial ponds carved out of mangrove swamps in Central America and Asia. The number of inshore trawlers like the *FJG* has plummeted, down from 32,000 a decade ago to 15,000 today, and many of those are

Vietnamese. I ask George how he can tell the difference between a Vietnamese and an American trawler: " 'Cause they're pointy, ugly, and they stink," he says.

"I'm surviving but I am not making a living," George intones from the wheel. "There's a difference. You work so close to the line now that you can lose your shirt in a minute. I don't feel happy that a lot of the older people are being put out of business. The government people tell us we should be happy 'cause we are the survivors, but I can't feel happy if the traditional people are out of work."

Thank God the shrimp are still here. But even their abundance is a cause for worry. One theory is that detritus-filled waters from the deteriorating marsh make more food available to juveniles in the estuaries. But that's short-term glory; the longer view is that the destruction of estuaries and wetlands will have to lead to an eventual crash.

And then there's the growing "dead zone" to consider. This swelling chunk of the Gulf of Mexico is media heaven, written about cataclysmically by publications large and small. Little can live in the dead zone—the size of New Jersey, in recent months, in the neighborhood of 8,000 square miles—a swath of Gulf from the river's mouth almost to Texas. Finfish and shrimp are lucky; they can escape. Little else can. Almost every spring, runoff from farms far away dumps nitrogen and phosphorous from fertilizers into the Gulf via the Mississippi. Manure from hog farms increases, too. What happens as a result is this: algae go berserk, growing blooms that go on for mile after mile. Then the microscopic organisms creating the blooms die by the zabillion and filter to the bottom. Other bacteria eat them down there, depleting the water of oxygen as they munch away and reproduce. The process, called eutrophication, happens in many bodies of water contaminated by organic wastes. Here, things are worse. The water on the Gulf's floor off the Mississippi's mouth tends to be more or less stagnant, weighed down by its salt content, trapped beneath the fresh water flowing above it. Any life unable to move fast dies fast. Mollusks can't get away, nor can starfish, worms, or crabs. They've all died. The floor of this part of the Gulf, under the best of circumstances a drab, muddy plain, is now a sterile, drab, muddy plain. Sometimes, it turns black with decaying bacteria and it smells of rot.

Dr. Nancy Rabelais is the recognized expert on the dead zone in this part of the world, although she and her husband, Eugene Turner, first studied the phenomenon as a team. He's the professor at Louisiana State University who has his students put minnows in estuaries to see which ones are eaten first. Dr. Rabelais was having a bad day when I visited her at LUMCON, which stands for Louisiana Universities Marine Consortium, a tower of laboratories overlooking the marsh. "I didn't name it 'the dead zone,'" she moaned. "The media began calling it the dead zone and then the media began saying that I call it that and now all the farmers and fertilizer people don't like me."

Her hair is long and straight, outlining her face sharply, and her face was not happy. "I have just been turned down two times for grants on hypoxia (the technical name for death by oxygen starvation). I am not feeling great right now," she explained. What she did say is significant, however. First, she told me that hypoxia is a problem in more than half the world's estuaries, although its extent in the Gulf is greater than anywhere else in the Western Hemisphere. Second, it is a fixable problem, not "long-term." Despite the media's excitement, the dead zone is only dead so long as nutrient overload exists. Third, fertilizer manufacturers don't want to believe a word of what she says.

This is where George and other shrimpers come in. The fertilizer people cite the fact that shrimp harvests appear to be holding up as evidence that the so-called dire effects of nutrient overloading is hogwash. In a peculiar way, that may be true. The fact is, George was making some pretty good hauls when I was out with him. He had been having a good season for the past several months. This may or may not have a lot to do with the dead zone. It's only a theory: as the dead zone enlarges, fish and shrimp flee the diminished oxygen. They tend to pile up at the perimeter, creating a thin band of high population density, very exploitable if a commercial fisherman knows what he's doing. Though George did not acknowledge the existence of the dead zone in his repertoire of problems, it could explain why his trawls were coming up full, at least when I was with him.[1]

It's the stories and the lore that goes with the years that lead me to call George the Prince of Shrimpers. "If anyone can find shrimp out here, it's George. He knows the bottom like he lives down there," says Alan Blanche, the deckhand. He leans toward me early one morning before dawn, a guy with a big open face and a wise mouth, as we pick through shrimp under spotlights, separating them from the twenty or thirty finfish species that come up in the nets. He whispers of George's expertise with religious certitude. He murmurs something about the tough and gentle love that George shows his fourteen-year-old son Chris, who works on the boat every vacation to help pay his school tuition. Right now, Chris is taking it easy, transfixed by some game show on the TV in the cabin, while his father keeps his eye on the depth finder next to the wheel, knowing that at a certain time of day or night, and depending on cloud cover or clear skies, shrimp will cluster along the bottom at a certain depth. George looks over his shoulder at Chris from time to time, swinging his gaze back to the horizon after assuring himself that whatever his son is watching is decent.

Chris works on the boat only if he agrees to use his earnings to pay for half the tuition at the Catholic charter school he attends. He ended his high-school freshman year the first in his class, with a 4.2 grade-point average, a number that his proud father likes to toss around.

Though George may be the Prince of Shrimpers, he's a beleaguered one. His stories, big and small, true or tilting the other way, show the ever-more-onerous burden shrimping imposes on him. A

[1] In January 2001, the federal Environmental Protection Agency announced that nine states along the Mississippi had agreed to reduce oxygen-depleting pollutants entering the river and would come up with a specific plan by late 2002. Worrisome, however, is that the agreement is just that, with no requirement for enforcement.

tug's angling toward the *FJG*'s port bow just now prompts one story. In the ever-leery relationship between the Gulf's commercial traffic and its pokey fishermen, the consensus among shrimpers and oystermen is that tugs are wild cards—can't be trusted— understandable when they're pushing steel barges loaded with coal or flammable chemicals that reduce their maneuverability to near zero. They tend to wander out of the channel leading to the Intracoastal Waterway and to New Orleans, tend to bully their way right through fishing boats.

At least that is what George says. I am pretty sure that any tug captain would say that it's the shrimpers who do the wandering. George got into a jam with his interpretation of a wandering tug once; it came close enough so that his trawl outrigger snarled into the side of a barge and George, as stubborn as commercial fisher-men get when their rights are intruded upon, was loath to give way. A crash looked inevitable.

"Hey," George yelled at whoever was in the cabin, "were you born an asshole or did you just become one?" The question had no effect. George's deckhand, a young kid, panicked and jumped off the stern. He drowned. "I just went home and sat on the bed and cried like a baby," George says.

Crying is alien to George, long ago eliminated from his behav-ior when he began working on his daddy's boat as a young boy. It's easy for him to think that those were the days when a man could make a proud living as a commercial fisherman. His daddy did, after all, despite a leg lost to diabetes. And his father's peers and his relatives, all descended from a figurehead who emigrated from Bosnia in the late eighteenth century, had made good livings as fishermen. No more, and family traditions are stumbling badly as a result.

A lot of bad things can happen out here where George likes to trawl in the St. Joe Channel off the Mississippi coast, and they seem to be getting worse. Sure, there are squalls and waterspouts that can drop out of the bottom of a thunderstorm like angels gone mad and hurtle across water, splintering everything in their path.

And there's debris on the bottom, like wrecks that can snag a trawling rig and spin the boat around 90 degrees, yanking a gunwale under with such force that you'd think a serpent had swallowed a net whole.

George warned me about that after I had been out with him for a day or so. He said that I should always walk leaning inward, especially along the narrow deck on either side of the cabin. "Those trawl cables hit you when you're in the water and you're gone and then the screw will suck you right under," he said ominously. Then he added thoughtfully: "Did you know that commercial fishing is the most dangerous occupation there is?" Sometimes George will become reflective, which fits his appearance nicely—with his bifocal wire rim glasses, dark wavy hair, and elegantly curved nose.

As a young man George thought of leaving the dangers. He made a stab at getting off the river, getting out of pulling rigs for shrimp and dredging for oysters, getting away from cramping his hands up in the wet heat of summer and the damp cold of winter. He went to college instead, unheard of among his family and unusual for most commercial fishermen. He graduated from Southeastern Louisiana State, in Hammond. Then he enrolled in Loyola University Law School in New Orleans. He said he felt like an alien every minute of the two years he spent there. Wearing a nice shirt every day was just strange to him. So maybe it was fortunate that his father came down with diabetes and eventually had a leg amputated.

George felt obligated to drop out of law school and return to the boat. "Besides, I had this feeling that I couldn't keep away from the water," he says.

George's bigger problem, looming bigger than physical danger, is increasing government intervention. He claims the government with its growing list of regulations could result in his financial ruin. TEDs (turtle excluder devices) are one nightmare. TEDs are a grill-like barricade that shrimpers are required to install in their nets to allow sea turtles to escape certain death if they are unlucky enough to get caught in a trawl. He snarls out that TEDs cost him

$20,000 in lost shrimp each year. "If it weren't for those things, I'd have paid off the house and my wife wouldn't have to work," he complains. Another source of financial doom is the government's effort to restore the marsh through freshwater diversions from the Mississippi. George says the silt that has come with the freshwater has smothered the oysters on much of the family's 1,650 acres of leased bottom.

During a sunset that lights the world with golden streamers from heaven, George winches his two trawls in for the third time that day. They rise from the water, suspended high by the outriggers on either side of the boat. The trawls drool bits of fish and shrimp and slobber great gelatinous hunks of jellyfish. The slimy webbing looks to be the jaw tissue of two green monsters from a hell below us. George curses, despite the bundle of sea life swaying in their stomachs. He points to the round metal grill near the tail of each net—the TEDs—which are supposed to allow passage of smaller life through them and into the tail while detaining larger objects. Turtles are supposed to slip to freedom through a flap in the net where the grill is lodged.

In reality, it doesn't work quite this way. A lot of shrimp get out through the flap, never mind turtles which George says he scarcely ever sees in these waters. George says these flaps could be his demise; he says that these devices could take down every shrimper in the region. He claims he loses about 20 percent of his catch because of them. And that adds up to various conclusions that he is quick to utter. One is that he loses money to save a turtle—the Kemp's ridley—which he insists should not be here in the first place. There is some truth to this. The species is not native to waters off the Louisiana and Mississippi coasts. It has been introduced to the northern Gulf from its accustomed waters off the Mexican coast in hopes of increasing its range. Since 1989, when inshore shrimpers were required to install TEDs, George says he has seen only four turtles in his nets. He is sure the government planted them in an effort to restock this part of the Gulf of Mexico.

"They ought to just leave them down in Mexico where they nest naturally on the beaches."[2]

The fact is that shrimpers and turtles have long been in conflict, especially since environmentalists got hold of the issue in the 1970s. In 1990, after years of research, the National Academy of Sciences concluded that as many as 50,000 sea turtles died each year in trawls. Ironically, shrimpers had contributed many of the statistics and much anecdotal information to researchers. Long before the official recognition, pressure had begun to mount on the National Marine Fisheries and the U.S. Fish and Wildlife Service to do something about the problem, especially flagrant when trawling seasons were underway and numbers of dead turtles turned up on beaches. The Florida tourism industry had an easy job convincing legislators to require TEDs and to enforce their use. Tourists don't like dead sea turtles. But in Louisiana and Texas, where beach tourism is not an economic powerhouse, TEDs pitted shrimpers against the government. Politicians sided with the shrimpers. Congressman Billy Tauzin, in fact, used the turtle issue to delay reauthorization of the Endangered Species Act.

The environmentalists won. On July 21, 1989, after a series of parries and thrusts, a court order was issued that TEDs were to be used on all trawls. Louisiana and Texas shrimpers would have none of it. The next thirty-six hours disproved the adage that people who make their living on their own are incapable of organizing. Directed by a man named Tee John Mialjevich, president of Concerned Shrimpers of America, trawlers by the hundreds blockaded ports along the Louisiana and Texas coasts. They clogged Belle Passe, which is the name of the mouth of Bayou LaFourche. Transportation to and from Port Fourchon was brought to a standstill, an event that didn't disturb Ted Falgout as much as you might

[2] The origin of the "planting" is interesting: as the controversy over TEDs heated up in the late 1980s, some shrimpers noted helicopters hovering over the northern Gulf from which large black objects fell. To some, this was conclusive evidence that the government was seeding the area with turtles. Actually, the black objects were dummy targets for search and rescue exercises by coast guard trainees.

think. In Louisiana, most people's hearts are with those who work the traditions of the landscape or seascape. Even though a man like Falgout runs a port for big gas and big oil, the struggles of shrimpers course through his blood.

The coast guard didn't like the blockade, of course, and threatened to bust it up. Nor did the National Marine Fisheries Service who, amid the growing publicity, began negotiating with the shrimpers. The Gulf shrimpers won, at least, a reprieve. The whole matter went back to the courts, which resulted in another blockade. But this one lacked spirit. The shrimpers knew that sentiment was building for the turtles. The National Marine Fisheries began enforcing TEDs regulations, throwing $8,000 fines at shrimpers caught without the devices. Now the shrimpers use TEDs and hate the government.[3]

The devices cause the loss of more than shrimp and money. TEDs seem to focus George's wrath; he credits them for much that is wrong in his life. George has bad dreams about catching derelict crab traps in his rigs. He dreams about how the mashed wiring clogs his rigs and tears at his nets. Why he has dreams over the crab traps I don't know; snagging does not happen only in his dreams; it's an everyday mishap and TEDs have nothing to do with it. Nevertheless, George has to stop everything and sew up the damage. He angers fast. "I say fuck the government. They don't have any idea how much time I lose having to sew up these holes." Over the past week, George had brought up almost two hundred orphaned crab traps. He piled them high on the bow of the *FJG* until he got to Bayou Caddy, where he unloads his shrimp. There, he tossed them ashore to rot in the marsh.

"I figure that surviving hardship is a test. If you make it, you're a better man." I think George uses this mantra to get by. It gives him a little pedestal, and when he manages to beat back a new gov-

[3] Much of the preceding discussion of TEDs and the blockades is drawn from an analysis of the controversy by Anthony V. Margavio and Craig J. Forsyth in their book, *Caught in the Net: The Conflict Between Shrimpers and Conservationists*, College Station, Texas: A & M University Press, 1996.

ernment regulation, he can get up on that pedestal and keep smiling at the world, at least for a little while.

At two A.M., under a star-sequined, ink blue sky, George winches in the trawls and swings them over the gunwales, just one more sequence in the timeless harvesting of the Gulf's floor. The nets hang dripping while he unfastens their tails to release their contents. The secrets of the Gulf tumble into plastic tubs on the deck, a panoply of riches—known by the pedestrian word as "bycatch"—which will soon go over the side, except for the shrimp. Above us, laughing gulls float like guiding spirits.

Grizzled Alan Blanche, barely awake from his few hours in the bunk, stumbles on deck looking like an apparition under the bright lights that George has directed on the work area. The routine of the catch is so etched into Alan's movements that he doesn't have to think. He hefts an overflowing tub to the saltbox. It is a big wooden container filled with brine. Its purpose is to separate the living from the dead and dying. Corpses and near corpses of the bycatch float to the surface. Crabs, catfish, and some flounder have survived, flopping among hundreds of dead fish on the surface. Shrimp sink to the bottom. Alan scoops the carnage over the side with a net on a pole. Mobs of gulls cackle wispily in the dark and dive for the tidbits. A pair of dolphins emerges in the boat's side wake and rolls in the easy luxury of the handout.

Then Alan digs deep into the saltbox for the recognized riches—the shrimp—and begins scooping them into plastic tubs. Chris helps, but his young arms can barely deal with the weight. Alan delights in exhibiting his biceps as he rotates the net back and forth and lifts it brimming with shrimp. He jokes to Chris that he should do the same work every once in a while so he could build himself up. As he heaves the overflowing tubs to the sorting box, a tattooed and sweat gleaming *Melissa* undulates on his right pec, while an angel with enshrouded breasts and thrust-back head seems to fly off his right shoulder.

Melissa was his first serious girl, he tells me when I ask, as we stand at the sorting box picking shrimp from all the other marine life that did not make it over the side from the salt box. Chris has heard a lot about Melissa and laughs when her name comes up and says something about her being psycho. Alan tells Chris to shut up and after all he has never even had a girlfriend. Chris sasses him back something in a continuation of hours, days and weeks of survival banter. I learn that Melissa broke up with Alan two weeks after he had her name imprinted on his breast. But Alan says all will soon be set right because he plans to have an elaborate design etched over the name to cover it up.

It's not without cause that Chris terms Melissa a psycho. After breaking up with Alan, she found a new boyfriend. She urged him to take after Alan with a baseball bat. The new boyfriend did, coming up behind Alan and splitting his head open. Shortly thereafter, he stole a car for which he was sent to prison.

You learn all sorts of things on a shrimp boat. There's something communal about picking shrimp; it's a process of purification, separating iron from ore, wheat from chaff. The work encourages the sharing of confessions and dreams, even under the rock music that George puts on from the cabin. I guess he thinks it will help us to work faster. Then there's something about being under the star-jeweled dark of a Gulf night, out of sight of land, rolling in swells, our bow splash as white as snow, at the mercy of weather that can change in a flicker. A primal nature releases our reserve; stories, dreams, and fears roll out with the swells.

Alan can't stop talking. He's a sweet twenty-six-year-old kid from the bayous who grew up with not much education, a lot of girl chasing, and a lot of fighting. He tells me about girls and fights while Chris snickers at my side. Then Alan changes the topic. He begins talking about food. In the soft thick night air, the boat moaning beneath our feet, the wind sweet-smelling, Alan confesses his real love—cooking. He reminisces about the three restaurants in New Orleans where he worked as a prep chef. He lists recipes for softshell crabs, shrimp, speckled trout, and redfish. I compliment him on last night's supper of fried pork chops in a sweet sauce. His face glows beneath a week's worth of stubble. He bets that if he were to open a Cajun restaurant in New York, he could make it big. When I tell him that there are already a number of Cajun restaurants in New York, he is rendered speechless for a few unusual minutes before he starts in on his oyster recipes. I ask him if he would ever leave the boat to go back to cooking. He looks at me startled. "Hell, no," he says. "This is a good life. I know what I'm doing out here. After a week of this, I have a pocket full of cash, a quarter of the money we get from the shrimp, and I go and buy stuff. Then, I'm ready to come back for more."[4]

You hear a lot about the waste in shrimping but it's hard to believe until you see it. Here's my species count from one pull (two trawls up from the bottom): croaker, channel mullet, crab, flounder, stingray, moonfish, puffer, silver eel, speckled trout, menhaden, gaff-top catfish, hardhead catfish, channel pollock, oyster, oyster fish, American eel, and, of course, shrimp. All the finfish, about 50 percent of the haul, are babies, no more than five inches long. Almost all die in the nets. George says it's always been like this ever since he has trawled, and those who think that fish stocks are being depleted because of trawling are just plain crazy.

By now, everything is dead except the crabs and catfish and oysters and an occasional flounder. The crabs always seem to survive. I fling whatever is living back into the Gulf (except the softshells and the oysters for the next meal). The dead go back into the Gulf, too, but I toss some morsels into the scuppers for the gulls and pelicans that line the outriggers in neat rows, waiting. The picked and washed shrimp go into flaming orange plastic buckets. Alan wrestles them into the hold and layers them in shaved ice. Then, he dives into a bunk for a couple of hours of sleep. He'll be back at the sorting table soon, just as the eastern sky dawns against a barricade of mighty thunderheads. It's been like this for generations.

---

[4] Despite this assertion, Alan eventually jumped ship, according to George, after he was promised a higher paying job in a shipyard. The job fell through and at this writing, Alan was trying to get his deckhand position back on the *FJG*.

• • •

That night he fries up a bunch of soft-shelled crabs out of the trawls the like of which I have never before tasted. We eat them in the cabin as rain pounds against the windows. A squall pushing fifty-mile-per-hour winds makes the *FJG* shiver all the more as she strains under the rigs. The film *Blood Sport* is on TV and enthralls us between munching on the crabs. In the early part of the film, a young Jean-Claude Van Damme rescues a Japanese boy—an immigrant—from American bullies who harass him as he walks home from school. As Van Damme leans protectively over the kid, I think of George's comments about the Vietnamese fishermen and wonder if he sees a connection between the boy, the Vietnamese, and his own family, immigrants from Bosnia. I don't ask George to travel down that road, however. I see no need to risk shutting him down and depriving myself from witnessing the Prince of Shrimpers at work.

I first met George at the wetlands conference at Nicholls State University in 1998, sponsored by Congressman Billy Tauzin. George sat high up in the auditorium, slightly apart from the officials and the engineers. He wore a nice clean shirt and freshly laundered blue jeans. He looked alert, with the same readiness that he had probably exuded at law school. During the Q's and A's, he swung into action, up on his feet, pointing out to Jack Caldwell, secretary of the Louisiana Department of Natural Resources, for example, that freshwater diversions along the Mississippi were ruining oyster leases, that promises had been made and broken to reduce the water flow, that commercial fishermen were being hounded out of business due to increasing government regulations. Was that what Louisiana wanted for its children, he wanted to know?

George is a known player to people who attend such meetings, which is a unique thing about Louisiana—the local environment binds people together from all walks, from scientists to fishermen, from hurricane forecasters to alligator hunters. They learn to tolerate each other, eventually to listen to each other. Caldwell obviously had met George before and knew enough to clamp him down before he got warmed up. "Yes, George, I am well aware of your concerns but this meeting is not focusing on those issues," he responded in a chilled haste that yanked me out of the South's diplomatic politeness and reminded me of the North's fondness for an agenda.

George is good at making his points in flagrant ways. One night on his shrimp boat, he told me about one of his methods. Just prior to meetings called to discuss Louisiana's coastal problems, ones that he knew Governor Foster was going to attend, he sent the governor a packet of shrimp and a card just as a reminder of who George Barisich was and that he stood for the future of the commercial fishermen of Louisiana. Now, Foster knows George. Once he went up to Foster, George told me, and said something to the effect of: "Governor, if you were a commercial fisherman like me and you had to deal with regulation after regulation, what would you do?" George says that Foster hunkered over him in silence, blinking a number of times. It must have been an interesting sight. George stands around 5' 6" and Foster, 6' 4". Then, a confused look crossed the governor's face, George said. Foster turned on his heel and walked away without saying a word.

The presumed success of such stonewalling, at least in George's mind, led him to found what he calls the United Commercial Fishermen's Association, a homespun lobbying effort, if ever there was one, to try to put some muscle into fishermen's complaints. He launched it in 1993, two years after the TEDs regulation went in. Now, he is fighting officials on a broad spectrum that includes new regulations against the use of strike nets, bycatch reduction devices, and freshwater diversions. He puts out a sporadic newsletter to alert his small membership of the extent of continuing encroachment by the government. Each spring, he sets up a booth at JazzFest where he grills redfish, boils up crabs and crawfish, gives away samples, and bends ears of anyone who will listen about the fate of Louisiana fishermen. He has even traveled to Seattle and to Washington, D.C. to gatherings held by commercial fishing organizations where he has sat on panels that vilify government intrusion.

George has even gone so far as to compose a ballad in defense of his compatriots: He found a singer named Anna White and burned part of it on a CD. Here's some of George's ballad:

The Government went down to Louisiana.
They were looking for an industry to kill.
They were in a bind, they were way behind
All the small farmers had already rolled down that hill.
They came across Tee John pulling his rigs and pickin' his shrimp.
The Government man jumped up on the picking table and said,
Boy, let me tell you what.
I bet you don't believe it but we say you're a turtle killer, son.
And if you don't put those T.E.D.s in your nets we're going to put you on
*   the run.*
Big Tee John was no fool, you see. He knew which way to go.
He called all the good fishermen to a meeting down in Thibodaux.
Listen friend you better patch all those holes and set your tickler right.
'Cause if you gotta put in them T.E.D.s, your catch will sure be light.
Tell all your friends and neighbors. Hell, tell your President, too.
Let him know just what it is he's trying to do.
'Cause if you win, you keep your boat, and house and kids and all.
But if you lose, you know you're going to fall.
The meeting room was packed that day.
It was such a sight to see.
Thousands of working people; fighting to stay free.
It's people like this that made our country proud, strong and free.
Now listen very carefully and open your eyes to see.
This is still America and I'll die
Before you take these rights from me.
Many people spoke that day.
And the meaning was all the same.
This is my life you're messing with.
This ain't no God damn game.

There's a bitterness out on the water now and a sadness, too, as the commercial fishing industry in almost all parts of the world suffers depletions, and in some areas, particularly the northern Atlantic, virtual commercial extinctions. While fishermen like George still bring in plenty of shrimp and oysters, and the menhaden fishery in the Gulf feeds the majority of the chickens that we consume, there's a general acknowledgment in management and scientific research circles that the bounty is at an end. But George sees only encroaching regulations, diminishing demands, and rising expenses. He's on the edge and mad as hell. One night, as his boat groaned along, he told me about some of his troubles with the law, about government people boarding him to check out this and that, about one agent in particular shadowing him. One day the agent boarded the *FJG*. George laid a revolver on the counter near the wheel and told the man that he knew damn well that he was obeying the law and if he had to put up with this garbage anymore, he would use that gun. The agent backed off, and George called the agent's supervisor to report the incident. The agent disappeared from George's life, a minor triumph. There's still a bitter residue in his mouth over the incident, which George likens to a poisoning.

One day, when I was on board, George had to shut down the engine for a while. One of his trawls had come up riddled with holes where an abandoned crab trap had lodged and chaffed against the TED. The midday sun blazed down on the boat, eerily quiet in the soupy waters. As the outriggers at horizontal yawed back and forth, trawls dripping overhead, the *FJG* wallowed in troughs like a piece of driftwood. Aimlessness, a loss of purpose, suddenly dominated. George's hands moved faster than my eyes could follow as he wove his needle through the shreds to make the repairs. The resulting knots looked like a schematic of dance steps. George said he learned the knot from his daddy. "No," he corrected himself, "it was my granddaddy who really taught me this."

I asked him if was going to teach Chris how to mend. "No," he shouldered off my question with a shrug of irritation. "He's not going to do this. He's not going to go into fishing." Then silence, the emptiness an acknowledgment by a father that he was the last of a line. There would be no more fishermen in George's family.

Chapter Six

Peanut's Hunt

Sidney Michel, or Peanut, the crawfisherman who showed me the Exxon-Duck Lake Oil Field, was born a preemie. Even when he got out of the incubator, he didn't grow. His parents took to calling him Peanut. Eventually, he began growing, and growing, and growing. When he reached 6' 6", he stopped. But everybody in Morgan City, where he lives, still calls him Peanut. Many of his relatives are tiny. They live in tiny homes clustered in a three- or four-block area up against the levee that holds back the Atchafalaya River. Peanut calls it his "bitty little neighborhood" and he lives in a tiny home there, too, with his wife, Tanya, who is much shorter than he.

While Peanut looms large beside his family, he also stands out for another entirely different reason. He has decided to make his living from the marsh, in contrast to so many people around him who have forsaken its unpredictability in favor of pay-by-the-hour jobs. Because of this, people in Morgan City tend to look up to him, to seek his advice about the movements of catfish, crawfish and cloud formations, geese and garfish. He brings back to them the lives of their forbears, lives that have lost intimacy. In their eyes, Peanut assumes mythic proportions, larger than his physical self, larger than their daily lives.

Peanut, who is only thirty-one, is a pioneer of sorts, because of his choice of livelihood. He has traveled back in time to pursue the same work of his near ancestors. Meanwhile, his peers and even older men have abandoned the marsh and swamps as the main source of their livelihoods. They work instead the oil platforms, oil and gas pipe yards, or jobs in the services that support the oil and gas behemoth—dredging, driving, hauling. The manual labor offers steady paychecks but quick layoffs when the price of oil drops too precipitously.

Peanut and I are going alligator hunting, one of the disappearing annual cycles of harvesting the land. September is alligator season. The marsh and swamps are still hot—steaming and luxuriant—and 'gators are still active, particularly the smaller ones, stocking up for the coming winter hibernation. The bigger 'gators—those over six feet—slow their feeding at the slightest chill from the north. When the temperature reaches 72 degrees, they stop completely. But not to worry; as Peanut winds his twenty-three-foot aluminum work boat down Big Wax Bayou in the Atchafalaya Basin, this day late in the month promises a sun that will raise heated streamers of air over the marsh. The warmth should lead 'gators to hunt during the coming night in preparation for hibernation.

Not many people depend on alligators to fill their bellies or wallets these days, but it's a way to make a chunk of money during the monthlong season. The state strictly regulates the harvest, allotting tags to the big marsh and swampland companies. Land managers then dole them out to their favorite fur trappers, whose main occupation becomes alligator hunting each September. The number of tags depends on the estimated number of alligators inhabiting their holdings. When a hunter kills an alligator, he cuts a hole through the tail and inserts a tag. Whoever buys the 'gator must then record the creature's vital statistics and return the tag to the state, as means of tracking the status of the population.

Most of the 1,800 or so hunters who kill the 25,000-plus alligators taken each year in Louisiana do so more for the thrill than for the decreasing financial benefits. The price for skins and meat has dropped by more than 60 percent over the past fifteen years as alligator farming has turned the prehistoric creatures into a cash crop. Nevertheless, at a price per foot ranging from $6 to $24—during the 1999 season, the bigger the 'gator, the higher the price—a hunter can clear between $5,000 and $10,000 in a season.

Tightly controlled hunting and a wildly successful stocking program have resulted in one of the greatest comeback stories ever for a species recently near extinction. In 1962, the state halted all hunting of alligators after years of exploitation that went back to the Civil War when they were killed without mercy to supply boot leather for Confederate troops. Hunting was not permitted again until 1972, and then in only one parish. Statewide hunting reopened in 1981. Now, the population has shot up to well over one million, a 'gator for every four Louisianans. As in Florida, where hunting is even more regulated and the population far higher, Louisiana alligators are increasingly making nuisances of themselves on golf courses and in subdivision canals.

Peanut takes alligator hunting seriously. The forty-five tags he has been given by the state to hunt the 140,000 acres of the Atchafalaya Delta Wildlife Management Area will contribute not insignificantly to his annual income, supplemented throughout the year by harvesting other creatures as the seasons turn—mullet, catfish, garfish—during the late fall and early winter, and crawfish in the spring and summer. In the winter, he traps, mostly nutria and raccoon and an occasional otter. By now in late September, he has almost filled his alligator quota; he has six more tags left.

Peanut's anachronistic choice of livelihood—living off the land—intrigued me. What was going on inside him that drove him back to his Cajun roots? For centuries, long before the 1755 diaspora from Nova Scotia, Peanut's ancestors had found strength, some would say survival, in the land itself—working it for crops or cattle, or harvesting its resources. Was Peanut's return genuine or merely a visit back to another time, convenienced by fast engines, ready markets, and the hundreds of amenities of contemporary life that most people no longer seem to be able to do without? I imagined that this trip and others I would be taking with Peanut, deep into the marsh and swamp, would give me an answer that I hoped would not be too superficial.

He had tried to be part of the modern world. After finishing high school, he went to work in the oil fields as a pipe fitter, coupling sections of drilling pipe on the platforms. But welding paid more, so he hung around the welders and they taught him how to weld until he got good enough to become a welder. "It was real interesting work," he says, "messing with different alloys every day to try to get a perfect weld."

But then he saw what happened to his father. In oil and gas for thirty-seven years, the day he turned fifty-five his boss came up to him and told him he was no longer needed. "I saw the future," says Peanut, "an' I didn't like it." He thought about going to college, "but I don't think my mind is set up jus' right for that kind of thinking. I didn't want to waste my parents' money. Besides, the marsh is what I love."

Actually, Peanut could earn far more money than he does if he ever decided to go into modeling or film. He's a casting director's dream. His teeth are white and even and he has eyes—hazel, like those of some cats—that fix on you and you don't want to let go of them. A bit of blond frosting tops his black hair, and when he's cruising down the bayous, standing, always standing, in the stern of his work boat, that hair blowing straight back, someone passing him might think he's from somewhere else much better known and frequented than the Louisiana marsh.

But people don't think that because most people are from around here and they know Peanut. He's been roaming the bayous, the marsh, and the swamp all his life. He learned everything he knows from his daddy and grandpa and uncles, all French-speaking Cajuns who, Peanut tells me, have always depended on picking crabs in the processing plants to see them through bad times. Peanut, like many younger people in the region, understands smatterings of French, but he doesn't speak it. Yet, like many younger Cajuns, he employs French grammatical constructions in his talk. For example: as we cruise down a pipeline canal, and pass a well-adorned hunting and fishing camp on a barge, Peanut offers that the owner had made "an abundance of money." This is followed by the interesting observation that the same person had "offered me to go hunting with him."

Occasional fishermen nose their craft up to Peanut's as we travel to his hunting grounds. There's a moment of silence, shifting of caps, wiping of brows. Then the Louisiana questions spoken in every bayou meeting across the state: "How many ya got, Peanut?" "What's the biggest ya got?" always rushed out in suspended excitement, the questioners focused on the results of the hunt as well as just passing the time of day. It's important in this isolated land to make contact with whoever else is around. There's no telling when people might need each other.

There's disappointment when Peanut tells them that we are only setting hooks. But questions are quickly reconfigured: "How many you got this season? What was your biggest?" Then the stories begin about the guy over in Lafitte who got a fourteen-footer or someone else on another bayou who killed three twelve-footers, all from the same little patch of muddy bayou bank. Eventually, they look at me, knowing I am not from these parts. They look at Peanut and he explains that he is "takin' a feller out to show him some alligator hunting." They nod in understanding, make a little more small talk, drink a soda, and roar off.

It's about an hour's boat trip to Peanut's hunting lease deep into the basin, where the Atchafalaya River begins to unravel like a rope and splay out into little bayous that meander to the Gulf. That makes some thirty miles through country of the rawest beauty I have ever encountered. It's primitive here. You don't know what strange plant life is going to explode around the next bend. We pass by a bay where great mushroom-shaped protuberances push up from the muck below. They are draped with heaps of dead matted vegetation, like hideous scabs. Eventually, these emanations will unfold into giant ferns, but at the moment of their birth, they are green monsters of uncertain origin. Even in maturity, their arms arc skyward, their immensity placing them, like the alligators we are going to hunt, in a world that has all but vanished.

Peanut is in no particular hurry to get to his hunting lease. He doesn't have to be. The outboard he uses allows him to take his time, if he wants to. It's a 140-horse Mercury, an experimental model, air-cooled, with four cylinders, just like a car engine, designed to run quiet and reduce pollution. Peanut has been given it to test out. He tells me that Mercury asked its dealer in Morgan City to find a fellow who depended on his outboard for his livelihood, who would put hard and steady hours on the motor, day in and day out, running it through mud, smashing it over vegetation, clogging its prop. Peanut was the only person around that the Mercury dealer could think of who used an outboard that way. So Peanut, the man who had returned to the past, got the engine of the future.

After he puts ten thousand hours on it, Mercury will reclaim it and tear it down to see where fatigue has set in. In the meantime, he has a piece of silver technology on the back of his boat that drives him down the bayous like a bullet. Peanut relates all this to me in a voice stoked with pride, not quite bursting out in exclamation.

It is a glorious morning, though at only 80 degrees, this is what people around here call a cold front passing through. Big Wax Bayou turns into the smaller Shell Island Bayou, its banks lined with elephant ears flopped in the water, then bull tongue behind them, then cut grass waving higher, and willow trees behind, a sequence of bright greens receding to the duller blue tones of the willows. Great blue herons rise ahead in irritated majesty, little blues and Louisiana herons move out faster and with less attitude, kingfishers skitter away, gyrating up and down. Mullet launch themselves like Poseidon missiles and flop back with a smack; garfish roil the water ahead, swishing their great tails on the surface.

Peanut suddenly veers off the main channel, cuts back the engine, and creeps into Grey's Bayou, a tiny opening in the immense green. He wants to check on a bald eagle's nest high in some moss-draped cypress trees. The nest one hundred feet up is as big as a tree house. An adult female swoops out of a neighboring cypress but there are no eaglets and that worries Peanut. He lingers in the bayou near the base of the nesting tree, his brow furrowed, telling me over and over again that "they was just here two days ago." Reluctantly, it seems, he distracts himself from his concern over the missing eaglets and points out what he calls "*grand voile*," water lilies, as big and round as a hula hoop. "That's what the old people used to call them. They got nuts in them after they flower— a big white flower—which makes the marsh smell so sweet. The old people used to eat them nuts all the time. I ate them when I was a kid but now I don' believe anyone knows what they are."

Back on Shell Island Bayou, Peanut yells that he sees an alligator ahead, his voice still full of boyish excitement. I don't see anything, but Peanut wants me to see what he sees, so he keeps pointing and saying helpful things like "Right there, right there, you can't miss it. See the head." Finally, I do see it through the sheer force of Peanut's exhortations—the snout and then nothing until the eye arches. The distance in inches between these two telltale signs pretty much dictates the length of the creature in feet. This one, gazing at us from the water at the edge of the bayou is small, only four feet or so.

A swimming alligator of any size does not look very prepossessing until you consider that the jaw, just underwater, operates through a muscular system that powers one of the strongest living crushing machines known, able to pulverize the bones of cow, horse, or human with a single snap. In fact, the strength of the muscles can inflict damage even to an alligator's own jaw. Edward A. McIlhenny, of the Tabasco Sauce family, discovered this in a strange experiment on Avery Island, only thirty miles or so from where we are now. McIlhenny spent much of his life poking among the hummocks and bayous that surround Avery Island, his family's oil and salt dome empire from where the world's production of the hot sauce springs. In his *Alligator's Life History*, published in 1949 and considered a masterpiece of crocodilian observation, he describes the fate of Frank, an eleven-foot alligator that he caught alive. Mister Ned—as McIlhenny was fondly known in the marsh—wanted to test the strength of the creature's jaw. This was before the days of animal rights. He and "his men," as he calls them, thrust "a flat piece of two-inch-thick steel" between Frank's open jaw. Frank snapped his mandibles shut with such force that his longest teeth were driven right through the top of the upper jawbone so they protruded from the skin. McIlhenny says that the holes healed over after he extracted the broken teeth with pliers.

The flip side of this enormous strength is that the alligator has little capacity to open its jaws if any pressure is exerted on them. In fact, a person with an average-size hand can easily hold a large alligator's jaws closed by grasping upper and lower mandibles at the front near the nostrils.

Peanut's excitement diminishes as soon as he assesses the size of the 'gator he has spotted. "I don' like to catch them that small," says Peanut. "They don' bring the money and you use up a tag if you keep 'em." But Peanut should have no trouble filling his quota tomorrow. As the day runs on, we see about thirty 'gators cruising the bayous, most treating our presence with considerable equanimity, even curiosity. The bigger they are, the more agitated Peanut becomes. Toward the end of the afternoon, we round a curve and come upon a sandy point containing three monsters of twelve feet or more. Peanut practically jumps out of the boat. "Did you see that, did you see that?" he yells, after they have slid off the bank and melted into the water. Of course I did, and for a moment I experience the horrific thought that what I had just witnessed couldn't be *real*. So frequently does this scene appear on television nature programs that *that* had become the reality for me. What I have just seen appeared an illusion, a virtual reality that in its rarity assumed subordination to television reality.

This is the way Peanut catches alligators. From a plastic bin, which becomes increasingly odoriferous as the day wears on, he takes a chicken leg and thigh, or a hunk of deer viscera, or a mangled blue-winged teal that he shot a few days ago and debreasted. The meat swelters and stinks in the afternoon sun. He chooses carefully, hoping to match each bait to the alligator he has in mind. He saves the blue-winged teal for the largest ones, affecting a personal connection between himself, who shot the bird, and the large 'gators, who swim around for us to admire. He jams a fishhook with the diameter of a Ping Pong ball through each morsel and suspends it via a light rope clothespinned to a pole angled into the mud on a bank so that the bait hangs about a foot above the water. The rope is about twenty-five feet long and he ties its other end to a stake he has driven into the mud. Blood and fats from the mess on the hook drip into the water and float with the current. 'Gators down the bayou eye us. When the scent hits them, they turn and swim toward the boat nonchalantly, and not particularly threateningly. But they won't make the fatal leap until after dark. We will come back the next morning.

On the way back to Morgan City, the sun turns the cypress swamp to the west shades of pinks and yellows. The sky fills with spoonbills making their way to a night roost. An out-of-place chill suddenly fills the secluded twists and turns in the bayous, drawing down thick curtains of mist that part and close as if ghosts were trying to see through them. An ibis rises from a mud bank. Peanut doesn't call it an ibis; he calls it a *bec à rouge*, a French nod toward its pinkish bill. "The old people used to eat those," he tells me again. "An' to tell you the truth, them old Cajins livin' off the land an' all would eat 'most anything." Are we among their ghosts here? A barred owl stares down at us from high on the jutting branch of a dead cypress. I comment on the pure beauty of the place. Peanut's eyes widen. "Yeah, yer right, sometimes we jus' don't appreciate what we got here until we think about it, but I sure wouldn't ever want to be away from this."

Alligators are far fiercer in myth than in reality. Unlike their crocodilian cousins, they are shy creatures, loath to attack humans. Edward McIlhenny gleefully recounts his experiences as a boy (circa 1880), swimming with 'gators in a bayou near Avery Island.

> . . . boy-like we always took great pleasure and not a little excitement in seeing how many 'gators we could call around us during our swim. We would attract them by imitating the barks and cries of dogs and by making loud popping noises with our lips, as these sounds seemed to arouse the 'gators' curiosity, and they would come swimming to us

from all directions. We had no fear of them and would swim around the big fellows, dive under them and sometimes treat them with great disrespect by bringing handfuls of mud from the bottom and "chunking" it in their eyes.

Perhaps it is because alligators are timid creatures, after all, that these boys had so little fear of them. They are not aggressive hunters, swallowing—they are incapable of chewing—whatever they come across or comes their way—ducks, frogs, turtles, smaller alligators, and their favorite, nutria. Dogs, too, they seem to like. Get people in a bayou community or a weekend camp out in the marsh talking about alligators and invariably a story is told about someone's dog whose barking on the dock or along a bayou bank ended suddenly in a terrific yelp. And nothing was ever seen of that dog again.

Even females with eggs are not particularly aggressive, less so than most incubating birds. Alligator ranchers locate nests—great heaps of dead vegetation—from the air, marking the spot with an orange flag on a pole that they drop from a hovering helicopter. Guided on the ground by the flag, they approach the nest in pairs, extremely cautiously. The female is concealed somewhere nearby, hidden in marsh grass. While one person rips into the nest, the other stands guard, armed with a heavy stick. Inevitably, the alligator charges, but usually without conviction. Experienced egg collectors merely jam the stick against her snout and the animal backs off. She may make several more halfhearted charges, but the stick against her nose eventually discourages her and she disappears.[1]

---

[1] While egg collecting may seem cruel, this and strict control of hunting are largely responsible for the resurgence of alligator populations in Louisiana. Ranchers, with the permission of the State Wildlife and Fisheries Department, are permitted to raid only selected nests, usually containing between seventy and one hundred eggs. About 17 percent of the resulting hatchlings are then returned to the marsh—tagged and tail-notched for easy age determination—but only after they have reached a minimum length of three feet, about 1 1/2 years old. This percentage is an estimate of the survival rate of fertile eggs and hatchlings in the wild. In other words, in natural conditions, over 80 percent of the eggs and young do not survive.

Despite the ease with which nests are robbed, stories abound of fiercely protective alligators charging people and boats near their nests. In Texas, a female with a nest close to a road is even reputed to have charged a tractor. Such hyperbole fits appearance, exploited to the hilt by swamp-tour operators who naturally highlight alligators as they shepherd expectant tourists on pontoon boats through swamps and along bayous. One of their gimmicks is to feed alligators along the way hunks of chicken stuck on the end of a pole.

"Here Boomer, here Boomer," an elderly woman who called herself Alligator Annie (now retired), sang out as she guided a group along mysterious bayou passages on a tour I took some years ago. "Now, where could that Boomer be. He always comes when I call him," she played up to her audience. "Oh, *there* he is. Good Boomer." Boomer was a twelve-foot 'gator swimming toward the boat, plump with chicken that he had been fed every warm day for a couple of years. Annie, a sweet person, knowledgeable about the local environment and anxious to educate, held out a quarter of a chicken on her pole and told the camera-clicking crowd to stand back. Boomer leapt, opened his great jaws, showed off his prodigious teeth, clamped down on the chicken, and dragged it under. The tourists were thrilled, myself included. We all went home vicariously terrified by the thought of what it would feel like to have those jaws clamp down on a leg or an arm.

While the mythology perpetuates itself through experiences offered by people like Alligator Annie, it is also based on hunting lore from the past. As the alligator population steadily decreased in the early decades of the twentieth century, hunters probed deeper in the marsh. So timid were the creatures, the only way to locate them was in their dens, which they excavated in the banks of a channel or under matted vegetation, creating a network of tunnels sometimes forty feet in length.

Those who hunted 'gators by searching out their holes were determined folk. They had to spend days in the marsh looking for signs of alligator. While they may have carried an axe or a rifle, and a shovel, it was two poles that were their most important equipment, each about twenty feet long, one tipped with an iron hook.

Upon discovering a den, a hunter proceeded to jam the blunt-ended pole through the overlying marsh until the alligator's back was struck. Then began the process of gingerly digging the creature out and having no idea what to expect. The muck cast aside, the 'gator remained hidden under the muddy water. The hooked pole was then poked about until estimated to be under the alligator's jaw. A mighty jerk was then administered with the hoped-for result that the 'gator would emerge and present itself either to a bullet or an axe blade.

It didn't always work this way, as recounted to me by an old fur trapper named Wilson Verret, a sort of spiritual ancestor of Peanut. Unlike Peanut, Wilson didn't have an outboard. He used only a pirogue—a narrow craft, originally a dugout but later made from cypress and plywood that could be poled into narrow channels. He and his wife, Azalea, lived far out in the marsh for months at a time, returning to their bayou community only to sell skins and purchase basics.

"Ah was huntin' one time with another fella and we hooked a 'gator an' began to pull," Wilson recounted one time when I was visiting him in his tarpaper marsh cabin. "He wouldn't budge. The other fella said it was a log, but ah knowed it was a big 'gator. We pulled an' pulled an' he wouldn't budge. Ah tol' him to go git us some help, an' another person came, but that 'gator wouldn't move. Then ah got ma shotgun an' loaded her with buckshot shell. Ah knowed he was too deep under the mud for the buckshot to git at him, but ah thought the noises would shake him. Ah was right. Ah held the barrel a couple of inches from the mud an' let 'er go. That 'gator came outta there like a waterspout, mud flyin' all over the place. Ah didn't have my rifle with me. Ah always killed my 'gators with a knife. After he slowed down a bit, ah jumped right into that hole up toward the haid. Ah knowed that tail woulda broke me in half. Well, as soon as ah could, ah stuck him the spot right back of the haid. That was it. He was daid."

Now, that was hunting.

Even *this* 'gator didn't do much to defend itself. Most of the battle was in Wilson's digging, poking, prodding, and telling the story. No matter how passive the creatures act, humans seem to have to give them a ferocity alien to their nature. Imparting them with such a reputation suits our nature, not theirs. We seem hardwired to assume that anything with a fierce facial expression (gorillas), sharp teeth (alligators and sharks), or that slithers or has too many legs (snakes and spiders) is dangerous. Perhaps all the stuffed alligators that mob tourist shops in the French Quarter—costumed, smiling, and cutely posed—are mainly a vain effort on our part to beseech them to depart from their alien world and join ours to make us feel more at ease. It is just as well to be wary, however, for some of these creatures, polar bears, grizzlies, and crocodiles (in contrast to alligators), *are* dangerous. We don't seem to be able to differentiate very well between being dangerous and looking dangerous.

Early the next morning, Peanut and I head into the marsh again, expecting to find alligators on the fifteen-or-so hooks set out the previous day. Peanut takes a shortcut down the wide Atchafalaya River, full of huge barge tows pulled or pushed by churning tugs. Their wakes make the going rough. Crewmen look down upon us, registering only minor interest; the scale of the river and its traffic has negated the inclination toward cordiality taken so much to heart on the smaller bayous.

We pass a serious dredge laden with technology, pumping sand from the bottom of the river to a low area on the bank. The river is so full of silt that it is in constant need of dredging to maintain the shipping channel upstream to Morgan City. The pumped spoil is creating hundreds of acres of new land each year. This, along with sandbars and islands that rise out of the river through natural deposition, make this part of the state unique. Here is the one place on the coast gaining turf, about which Peanut comments that the world to him seems "all topsy-turvy, what with up north the world is disappearing with mud slides and whatnot and here the water is turning to land."

Peanut's young wife, Tanya, is along this time, curled on the little bow deck, her pale skin ready to be roasted by the warming sun that is quickly dispelling the night's uncharacteristic chill. She keeps peering into the thick vegetation on the banks as if she were looking for something she might have misplaced. Then I understand that she is just mesmerized by the mystery of all those dark places between the cypress trunks and the endlessness of little bayous and canals that come off the one we are now skimming down, creating a maze that beckons the soul.

A little further on, any sense of mystery is dispelled by the sight of horrifically obvious nutria, hundreds of them on a half-mile stretch along the shore, fixedly feeding on plants—bull tongue, alligator weed, pickerel weed—clear-cutting a wide swath of vegetation. Their coats are mud-matted and dank, their great orange teeth gnash against stem after stem, tearing away blankets of green down to the marsh's muddy scalp in their frenzied quest for food. Robotic in their determination, oblivious of us, an underworld god must have ordered them to destroy the earth's luxuriance. Tanya eyes them with a cold, assessing eye. Then she turns to Peanut. "Peanut, can't you get the trapping lease here? They's jus' so many of them." Peanut tells her that the landholder has long leased it to a trapper who, he opines, is not doing a very effective job. Tanya resumes her staring at the animals, now with sparks flickering in her eyes.

Whenever Tanya can get away from her job as a veterinary technician and dog trainer, she joins Peanut on his forays. Two weeks after they married, Peanut informed her that he was going to leave welding and spend all his time working the marsh. "I said, 'Sure, you go ahead; I know how much you love it.'" She is proud she told her husband this, that it was okay with her for him to quit his job; not many brides would have been as generous. But she knew Peanut well enough to know he would make a living and it would give her a chance to spend time in the bayous, too.

"My daddy always wanted a son," she says, "but he got three girls instead. He was always bringing us out here, and I love it."

Out toward the Atchafalaya River's mouth, where it broadens out and its exhausted current dumps sediment that creates a maze of little bars and islands, we come to the first hook. "Bait's down," Peanut says excitedly, meaning that something has yanked the dripping chicken or deer heart or duck carcass off its clothespin. It's a small 'gator, up in the bull-tongue plants and when Peanut pulls on the rope, it comes out, initially like a sunken log and then, like a spitting cat, thrashing the water into a white fury and smacking the gunwale with its tail. Peanut's in a quandary. He's got six tags to fill and this is the first hook, but the 'gator is only a four-footer—less than two years old. He doesn't know if bigger 'gators are on other hooks. He decides to leave this one hooked. Depending on what lies ahead, he will either return to collect the creature or to cut it loose.

There's a bit of tension in the boat now. Peanut had set hooks out only where we had seen *big* 'gators swimming the day before. But a little one had taken the bait and that makes no sense to him, an affront to his sense of logic. The second hook has a little one on it, too. Peanut cuts it loose, muttering to himself that right here at this very spot we had seen a ten-foot 'gator yesterday. And where was it and why didn't it want that teal and this makes no sense at all. Where did this little one come from, anyway? Reluctantly, he theorizes that the big 'gator is a female and this was one of her offspring still hanging around the nest, probably a correct assumption, given the highly developed nurturing behavior of female 'gators.

The next half-dozen hooks remain untouched. As the bad news gets worse, Tanya keeps glancing back at Peanut, upright and enormous in the stern, face sour. Bewilderment clouds her smooth, cheerleader face. This is not what Peanut had told her was going to happen today. He had said to her, as he had to me, that we were going to kill six big alligators today. They were right out here yesterday waiting for the hook. It might be running through Tanya's mind at this moment that Peanut got it all wrong, that her marsh-loving husband actually cannot read an alligator's mind, or it might be that she is worried about how Peanut is going to behave. She looks back at me, too, from time to time, with a concerned expression, wondering how all these empty hooks are affecting me.

Peanut returns her look with a scowl of frustration rooted in money and pride. Dollars are riding on the six 'gators he planned on getting this morning, at least six hundred dollars, paid to him by a processor who would send these 'gators the way of many dead Louisiana alligators—up to Catahoula for skinning and butchering, and then the meat on to restaurants and the hides to Paris for handbags and shoes.

"Bait's down," Peanut spits out as we approach the next angled stave. The attached rope leads out into the water, a hopeful sign. Large alligators tend to spend time in deep water when hooked, rather than up on the bank. Peanut pulls on the rope and the water explodes, jaws clack and the boat shudders each time the tail crashes against the gunwale. Peanut strains to pull the head closer, but is unable to even hold on. He lets go. There's silence.

"He's a good one," he says, his voice soft with anticipation, "What ya think, nine feet, mebe nine and a half?" Tanya thinks Peanut is off. Eight and a half max, she says. Suddenly she's holding a revolver, a .38 Special Smith and Wesson that has materialized from somewhere. She looks much too innocent to be wielding the weapon but does so with obvious familiarity, barrel angled away from her body. She walks toward the stern and Peanut pulls on the rope again. The water roils mud from the bottom and the 'gator's head emerges, thrashing, then still. Tanya takes aim from four feet away and pulls the trigger. Blood geysers from a tiny hole toward the rear of the head. The animal contorts and freezes.

"They always tighten up when you shoot 'em," Peanut explains, gathering a splayed foot and yanking his prey upside down so the blood spews out faster. Tanya has returned to the bow, empties the chamber, and returns the pistol to a sheepskin case, giving it a pat. Peanut hauls the carcass aboard and stretches it out in the bottom of the boat. The 'gator measures barely eight feet, but Peanut is smiling as he puts a slice into the tail and inserts a white plastic tag with a number on it.

The animal is magnificent, as strange and otherworldly as the nutria, still armored for battles that ended millions of years ago, but with underbelly skin as soft as a baby's bottom. Even in death, the 'gator walks, moving its hind feet back and forth along the bottom of the boat, skewing its tail back and forth as if in a dreamy trance.

As we travel to the next hooks, Tanya keeps looking back at the alligator on the floorboards, then looking at me and then at my feet planted six inches from the dead 'gator's head, then at the 'gator. She's got her worried expression again. As if I am not there, she and Peanut begin talking about dead alligators that come back to life. "I shot a 'gator once," Peanut says, "an' after about an hour I noticed that he was blinkin' his eyes at me. I knew something was not right. The next thing I know he was takin' a bite out of the ice chest." I shift away from the 'gator.

Tanya listens patiently and then tells a story about a "dead" 'gator in the bottom of a boat getting up on its four feet and charging her uncle who was in the bow. "He shot it again as it was comin' for him an' the bullet hit his head plate and ricocheted back and grazed my daddy right below the eye." Shooting alligators in the head, even at close range, is not as easy as it might seem and I am filled with new admiration for Tanya's aim. The bony structure atop the head, which protects the small brain case, is thick enough to stop a bullet even from a high-powered rifle. Entrance to the brain can be gained, however, just aft, at the beginning of the neck. But a bullet fired into this softer muscle mass must be angled toward the front of the skull so it can travel under the bone and into the brain. While a 'gator can be knocked unconscious by a poorly aimed bullet, it will be roaring with anger when it comes to. Hence, Tanya's concern.

Five more hooks, three more alligators, each only six feet. Then a seven-footer. Now, five alligators lie in the boat, in various poses of death. An hour has gone by since we got the first one, and it is still trying to push itself along with its hind legs. But the movements are weaker and jerky. Tanya looks less worried, but her husband is beginning to mutter to himself again. When we round a bend and he sees bait hanging where a twelve-footer was cruising yesterday, he explodes. And when the very alligator that should be on the hook tauntingly surfaces nearby, he turns to Tanya: "Give me that pistol, honey," he says.

"Peeenut," she cries, drawing out the first syllable, "You cain't kill a big 'gator like that with this little gun." But she hands it to him anyway. The boat and the 'gator have drifted about fifteen feet apart by now. He takes aim, using both hands, and shoots. The water roils, the 'gator goes under. Peanut looks disgusted until, unexpectedly, the creature emerges again, clinging to the shore but looking undamaged. The boat in the meantime has drifted toward the middle of the bayou, widening the gap to twenty feet. Peanut pulls off three shots and the 'gator disappears.

"Peanut, there a boat coming down the bayou," Tanya says.

"I don' care, there isn't anythin' wrong with shootin' 'gators long as I got tags for 'em." He raises the pistol again as the 'gator rises, as if mocking him, and fires two rounds.

"Peanut," Tanya snaps, "It's the feds." Three uniformed men are in the boat, angling toward us, not feds but state Wildlife and Fisheries Department wardens patrolling the bayous to bust illegal duck hunters or anyone else who needs busting. They nose up and all say "good morning" real quiet and polite, one defined by blue reflector sunglasses, one by mirror reflectors, and the third by dead eyes that reflect nothing at all. Automatics lounge on hips, and pitch-black life preservers against standard-issue, mustard green uniforms impart a paramilitary sinister look.

The five 'gators in Peanut's boat are assessed; the tags in the tails of each are duly noted. The wardens can go no further in that direction. Peanut tells them he's trying to shoot an alligator that got away with one of his hooks, revealing his nervousness about this eerily subdued confrontation. He has chucked the revolver in a plastic container filled with empty Coke cans, old rope, and rags. The three men nod simultaneously, and say nothing. The cicadas in the willows on the bank are operating at a high decibel that rattles eardrums. The rasp escalates into a scream, the longer the human silence endures, shaking the horizon and torturing the world.

"We wanted to get a big 'gator for him," Peanut says finally, jerking a thumb toward me. "He's from New York."

Blue Reflector rotates toward me, registering interest in this information through a lift of his brows. He tells me he has just come back from there. I ask him where. He says New York. I ask where in New York. He names a place that sounds like Sacketts Harbor. I ask him where that is. He tells me it's in New York. I ask him where in New York, again. He's getting irritated now. It's somewhere on Lake Ontario, he tells me, and stares at me for a while through his blue lenses, his brow furrowed. I tell him that I don't know the area but offer that it must be pretty beautiful. "It's okay," he says. I observe that he might have seen the Adirondacks, thinking that that would be a sight for a Louisianan. "Yeah, I saw them, they're okay," is all he says.

He rotates toward Peanut. "You got three life preservers on this boat?" he asks conversationally, seeming to be far more interested in the big 'gator inching closer to us, the beneficiary either of a very thick skull or of Peanut's poor aim.

"Sure we got 'em," says Peanut casually, then he turns toward the 'gator and he raises his voice in exaggerated excitement. "Look at the size of that 'gator! That's the one I was trying to shoot!" We all stare at the 'gator for a while and listen to the cicadas.

"Well, could we take a look at them?" This from Mirror Reflector, having a high old time taking it easy on one of the boat's cushioned seats, his feet up on the steering console.

"Tanya, honey. Reach into that compartment there and pull out those preservers, will ya?" Tanya grabs one, then another one and then says, "Peanut, that's all there is."

Peanut is dumbfounded. He could have "swore" that there were three there this morning.

"Mr. Michel, sir. Do you have a driving license, sir?" Dead Eyes stirs awake, but keeps his gaze fixed on the bayou's far shore, now a hint of smile playing across his face.

Peanut searches his pockets but guesses he "plum" forgot it.

"Mr. Michel, sir," Dead Eyes picks up again, "I'm goin' to write you out a violation, sir, for the lack of required number of life jackets, meaning, sir, that if there are three people on this here boat, you have gotta have three life jackets." He's sitting next to Mirror Reflector, comfy-looking on his cushion, his hands resting on the wheel. Dead Eyes doesn't move for a full minute, then with reptilian casualness, he reaches under the console and withdraws a citation booklet.

From foreign parts, I am convinced that there is something very wrong with this picture. I guess something in my face translates that. As Dead Eyes writes out the ticket, Blue Reflector leans toward me a little. "This may seem surprising," he confides, "but down here we don't cut any slack when it comes to life preservers. Anything can happen out here." He sweeps his arm across the empty bayou and endless marsh beyond to indicate the "here," raising a few egrets and a great blue heron as he does so.

It's the pistol, stupid, I want to yell out. Is it legal for these people to be waving a pistol around at alligators? Don't they have to have a permit like up north? But I say nothing.

Dead Eyes proffers the ticket to Peanut who thanks him rather profusely without a tinge of sarcasm. The boats separate and the wardens head back up the bayou, looking back at us. Peanut reaches into the bucket for the revolver, takes a bead on the large alligator who by now has circled the boat about five times, and pulls off a couple of shots. The creature splashes and descends; Peanut curses that he missed again, and revs the skiff's motor. As we head back to the first hook, where the smaller alligator awaits a bullet to make up a total of six for the day, the big one surfaces and seems to be trying to swim after us but is soon lost in our wake.

The atmosphere has changed precipitously as it always does under the emasculating effects of police authority. Peanut gripes, of course, about the ridiculousness of the charge and says friends in Wildlife and Fisheries will get him out of the fifty-dollar fine. When we get to the little alligator, Peanut starts in again complaining about the poor appetite of the big ones. Then he is suddenly struck by an idea. Why the big 'gators didn't behave as he had predicted they would is because the "cold front" that passed through last night slowed up their feeding. Somehow, the realization that a natural phenomenon was responsible makes him feel better.

Back in Morgan City, we take the six 'gators to the buyer, hauling them in the now-trailered boat behind Peanut's Chevy Suburban, which he recently bought used from a friend. No skinning out is necessary anymore, the labor-intensive task after the glory of the hunt, a time when the victorious used to swap stories and etch their exploits into a society's conscience. Now, dealers purchase whole 'gators and hire someone else to do the skinning, a cost passed on to the consumer but not providing the hunter any financial gain. A young man named Travis climbs into the boat and swings each corpse to the ground like so many sacks of flour. They are measured, the tags removed, the numbers and vital statistics recorded. Then, they are slung up to a refrigerated trailer truck stuffed with hundreds of other carcasses awaiting a long journey from the bayous. Peanut is strangely quiet during this rapid transition from marsh to commerce, staring after his 'gators even after the truck doors have slammed shut. He stares vacantly at the check he is handed, something under $700, before pocketing it, the necessary money deflated in meaning now that the hunt has ended in no more than a business transaction, only a job, after all. Tanya says nothing. They both mount the Suburban and head home, away from the marsh.

Chapter Seven

The Marsh Eaters

Noel Kinler and I are sitting in an airboat in the middle of the Salvador Wildlife Management Area, a 31,000-acre marsh a half hour outside of New Orleans. In fact, from my cushioned seat on a strutwork of metal tubing built up from the boat's floor, I can plainly see the Shell and the Plaza Towers and the Energy Center in New Orleans, etched white thirty miles across the flatness. Straight up, a bald eagle cruises the deep blue sky, the December sun reflecting its glistening head with the purity of new snow. Recently arrived from the north, it may be refurbishing its nest for the approaching breeding season.

Directly below, the marsh looks, actually, like pretty solid ground, something like a soggy soccer field that has seen too many scrimmages, mud-tossed and messy between the vegetation. Noel tells me that despite the illusion of firmness, if I were to jump out of the boat, I would be up to my armpits in muck. This is a floating freshwater marsh; what Noel calls, in management bureaucratese, "an inland freshwater marsh situation" but what most people in South Louisiana know as a *flotant*, French for "floating marsh." The vegetation here springs from an otherworldly mattress of compacted roots which itself rests on an "aqueous solution," as Noel terms it, of a thickness you don't want to explore. Some roots penetrate eight feet down into this goo. This is the kind of marsh that a hurricane tends to roll up in tight rolls like those you see at a suburban turf farm.

The silence around us is of cathedral immensity, now that Noel has cut the plane engine, whose scream could permanently damage one's hearing within minutes without ear protectors. We are having a chat. Noel, purse-lipped and monosyllabic when I met him at the boat ramp an hour ago, is loosening up under my babble of questions. He has slung one leg over a pipe armrest, turned himself toward me. Behind his sunglass-obscured eyes, I see glowing the light that comes with the sharing of meaningful information. Maybe the wind freshening off Lake Salvador helps, pushing the sweet scents of marsh and water our way. It's hard not to relax under the spell of those perfumes. Nutria, along with alligators, are his research and management passion. He talks volubly about both, but mostly about nutria.

There's movement below us—nutria—hundreds of them all over the place. They are the reason we are here, to see what they have done to this great marsh lawn. They have mowed the hell out of it. One of Noel's jobs with the state Department of Wildlife and Fisheries is to keep his eye on the proliferation of these creatures. Along with people, the nutria are doing an impressive job destroying the marsh. Noel figures that across Louisiana nutria have gnawed around 100,000 acres of marsh down to root ends.

They're at work out there, sodden humps of fur with skinny tails longer than their bodies, chewing in their nonstop feeding frenzy, heads down in the muck, giant orange teeth severing the tender roots of pennywort and spike rush. Killdeer scream and wheel in weird rejoicing, accompanying Noel's staccato bursts of information—the animals, their commingling with humans, the plants they eat, the marsh they destroy.

Like people, nutria breed through the seasons. Sexually mature at six months, a female spends virtually her entire life pregnant. Two days after producing a litter—typically four to five kits—she is ready and willing to accept a male's advances, and no particular male at that, thank you. Meanwhile, the babies, born fully furred and teethed, eat grass and roots from birth—one-quarter of their body weight every day, about three pounds of marsh biomass daily.

Mercifully, evolution has imposed some checks on such fertility, but minor ones. One is a relatively long gestation time— around 130 days. Still, a female can bear three litters every fourteen months or so, potentially fifteen new marsh decimators. So, if one female gives birth to, say, six females in a year, those six may produce thirty-six females by the time they are 1½ years old. And those thirty-six may produce 216 females within another 1½ years. Multiply this out among the fifty or so nutria that we are looking at, which is half the number out there munching away. A lot of damage can be wrought in a very short time.

If I could hitch a ride on the back of the eagle that is still soaring above us, I would get a very good look at the extent of the damage. The messed-up marsh that we are sitting on is only the beginning. I would see, as would anyone fortunate enough to soar with an eagle, a sprinkling of little round ponds across the marsh. They would appear like so many sequins scattered on a green shawl. Many of these were created by the nutria's voraciousness. As time goes on, through the work of winds and waves, the little ponds enlarge in a relentless carving away of the marsh.

"Roots hold all this soil together," Noel explains, adopting a rather philosophical tone, quite opposed to the machine-gun fire of facts and figures that he has just concluded. "If there's no root stock, you get erosion." These little beasties, then, through their prodigious reproductive capacity, possess in their teeth and gut the ability to literally sever the marsh from its roots.

In all fairness, the animals have not wrought devastation unaided. The subsidence endemic to South Louisiana eases the way. As the land sinks, the water creeps over it, first freshwater but eventually brackish, then salt, as the sinking progresses, killing off most of the thirty-plus species of freshwater marsh vegetation. But it does worse. While the vegetation needs water for life, it also needs dry land to begin life. Seeds of few grass species germinate underwater, and even those that have germinated will fail to send forth shoots if they are beneath the surface. So, if nutria have a go at the roots of a sinking marsh that receives no replenishment of silt or nutrients, and no regeneration because seeds cannot germinate, the animals are, in essence, clinching the extermination of an already doomed marsh.

And that is happening right where we are. I can see a nutria a few feet from the boat enjoying a freshly flushed root; its fellows are all up to the same pursuit. They don't seem at all alarmed by our presence; in fact, the nearest one appears as if entirely lost in its world of mud and fodder, staring at us with eyes like reflecting orbs as it works on a dripping root. Noel reaches out over the gunwale; the animal bares its formidable teeth at him but then buries its head under a tangle of vegetation, too easy a comparison to the proverbial ostrich. He picks the thing up by its naked tail and it twists around. Unable to maneuver to take a chunk out of Noel, it gives up and dangles in a semblance of complacency. When he releases it, it skitters away, zigzags for ten yards, and buries its head in another tangle to rid itself, I suppose, of the problem our presence has presented to its limited mental facilities. Noel, a little disappointed by this behavior, shakes his head and comments on the generally agreed-upon lack of intelligence of these animals.

I find their teeth fascinating. They could inflict serious damage in a flash, yes, but of greater intrigue is their color. It occurs to me that those flashing orange teeth could affect the national economy, due to the voraciousness of their owners. I say as much to Noel. He looks at me with a pleased expression and produces a rare smile. He turns philosophical again. "What does not register too often is the consequence of diminished plant biomass. It affects everything in the food chain." Less of everything is the specific effect; less vegetation results in less detritus—the bits and pieces of organic debris that crawfish savor and that nourish the myriad shrimp larvae in the marsh's crooks and crannies. All the little fish that turn up in shrimp trawls depend, for a start, on the marsh that is being destroyed beneath my feet. Without the steady flow of the stuff of a marsh's life cycle toward the Gulf, less will result—fewer waterfowl, crabs, alligators, shrimp, eventually, even nutria. It will take a long time for this to happen, but the process has begun.

Bayous remain the north-south transportation link in South Louisiana and villages strung along their levees are still dependent on an unpredictable fishing industry. *(Credit: Christopher Hallowell)*

Thousands of miles of oil and gas industry canals have been dredged through the marsh, creating, in effect, an artificial environment that is now seen as natural. *(Credit: Philip Gould)*

The Mississippi River Gulf Outlet draws a seventy-six-mile straight line through the marsh. It is responsible for destroying thousands of acres of wetlands. *(Courtesy: Port of New Orleans)*

As the land sinks and the gulf rises, sights like this are increasingly common. *(Credit: Willa Zakin)*

One of Louisiana's oldest oil fields, the Leeville Oil Field, shown here in 1938, is nearing exhaustion. The miles of canals dredged out of the marsh to provide access for equipment and pipe will remain long after the oil is gone. *(Credit: Fonville Winans, courtesy of fonvillewinans.com)*

Although thousands of in-shore shrimp boats once trawled shallow waters, their numbers are now diminishing. *(Credit: Christopher Hallowell)*

Three million-plus
nutria have devoured
over 100,000 acres of
Louisiana's wetlands.
(Courtesy: Donald Davis)

The flooding in New Orleans as a result of Hurricane Betsy in 1965,
here looking north over I-10 in construction, may have been a taste of
what is to come. *(Courtesy: U.S. Army Corps of Engineers)*

Severe flooding, such as this one in 1871 showing Common Street in downtown New Orleans, were fairly common until an adequate pumping system was installed in the early 20th century. *(Courtesy: The Historic New Orleans Collection)*

Buttressing and repairing existing levees along the Mississippi remains today an ongoing process begun by French settlers in the 17th century. *(Courtesy Donald Davis)*

The 1843 crevasse on Chim's Plantation in West Baton Rouge. *(Courtesy Donald Davis)*

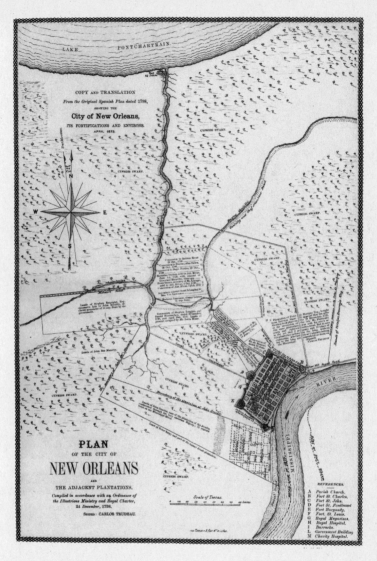

This 1798 plan of New Orleans shows not only Bayou St. John's attraction as a transportation link but the future city's vulnerability both from the Mississippi and Lake Pontchartrain. *(Courtesy The Historic New Orleans Collection)*

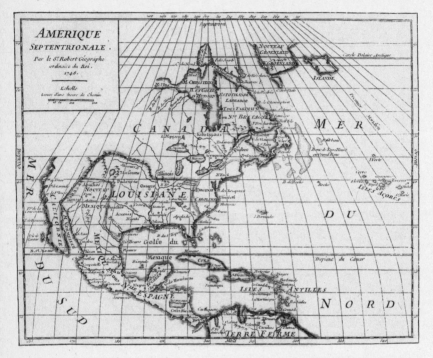

By the middle of the 17th century, the territory called *Louisiane* obviously occupied a large place in the minds of the French court. *(Courtesy The Historic New Orleans Collection)*

• • •

Nutria are accidents in the Louisiana marsh. First introduced by Edward A. McIlhenny, who kept some on Avery Island, their rout of the surrounding landscape may be causing him some anguish in his grave, for he was a lover of the marsh if ever there was one. His descendants still live on the island in the same stone mansion that his father, a Yankee from Rahway, New Jersey—and the inventor of Tabasco Sauce—built before the Civil War, when McIlhenny came south to woo and wed one of the Avery daughters.

Mister Ned, as the local people called this man's son—a wanderer of the marsh and companion to trappers and hunters—was a remarkable gentleman scientist of the nineteenth century. It is fortunate that he never had to work for a living; instead, he spent years observing, collecting, and writing. His *Alligator's Life History*—his best known work—is a minor effort compared to his treatises on botanical subjects. Specimens of his lifelong fascination with exotic flora still thrive in the island's lavish gardens—papyrus from Egypt, papaya from South America, soap trees from France, camellias from India, and Wasi orange trees from Japan.

Nutria interested him, too. He wanted some. It was a curious desire, given his inclination toward flora. Though nutria had been eradicated from their native South America, with the exception of Argentina where they were trapped, they were known entities in the Northern Hemisphere through introductions. In the late nineteenth century, a Frenchman farmed them in the Loire valley for their fur. The idea caught on as a cottage industry, pelts being sent to Berlin and transformed into coats. Around the same time, a nutria ranch was started in Elizabeth City, California, and backyard nutria farms were spawned throughout the Midwest and along the eastern seaboard with promises of fortunes to be made. Of course, many of the animals escaped and the operations fell apart. Nutria that didn't escape were often released when would-be ranchers began to calculate the work of killing, skinning, stretching, and shipping pelts. In California, the animals devoured so many crops that farmers went on nighttime shooting sprees, eventually exterminating them.

In 1938, McIlhenny received twenty animals from a New Jersey nutria farm and put them in a pen. Two years later, there were 150 of them in enclosures, all descendants of the original twenty. The experiment, if that is what it was, ended suddenly when a hurricane ripped through the marsh. The pens were torn apart, scattering the animals. Those that survived are the probable ancestors of the three million plus that are now devouring the marsh.

Initially, their presence was appreciated. They ate water hyacinth voraciously, an alien species which, since its introduction in 1884 at the Cotton States Exposition in New Orleans, had proliferated and begun to clog waterways. McIlhenny's escapees and their descendants were put to use, rounded up in live traps, and then released all over the state to keep down the hyacinth, just as profligate in its way as nutria were in theirs. While the effort had little permanent effect on hyacinth, the dissemination of nutria helped set in motion the weakening of the marsh's fabric.

The animals were not content with just hyacinth. They attacked sugarcane and rice fields. They burrowed into levees; roads and fields flooded. They harassed muskrat, the dominant fur animal in Louisiana, by overrunning their marsh terrain and consuming their food supply of grasses. Something had to be done. Not until 1949 was their potential as a fur animal hit upon as a solution. Then trappers, tradition-bound to muskrat, had to be swayed to go after nutria. Publicity campaigns and incentives offered by the Louisiana Department of Wildlife and Fisheries eventually took hold.

By 1962, over a million pelts were sent to Germany for processing, and nutria quickly replaced muskrat as the mainstay of the Louisiana trapping industry. By 1976, over 1.8 million animals were killed for their pelts. Nutria no longer were a problem in the sugarcane and rice fields. And trappers were happy, receiving, by 1980, just over eight dollars per pelt and earning almost $16 million.[1]

---

[1] Older trappers accustomed to muskrat were not all that happy, however, as nutria overran the muskrat population. Though the price of a nutria pelt rose to over eight dollars, trappers received around $4.50 for a prime muskrat pelt. In the time it took for a trapper to skin, scrape, and stretch one nutria pelt, he could process three muskrat pelts. He thus lost $5.50 for every nutria, provided there were muskrat to trap.

Anti-fur activists brought this nice arrangement to a jarring halt. After they began spray-painting fur-draped models on New York's fashion runways, the price of nutria dropped to $2.64 per pelt. The number of animals trapped plunged by over 50 percent, and the acres of marsh destroyed by the once-again burgeoning population began to shoot up.

Protestors may not have fully considered the effect of their actions. Not only did they weaken the fabric of the marsh; the protests and a sympathetic media and public exposed the animals to more suffering than leghold traps ever could. So prolific are nutria that their population explosions can end in either mass starvation or mass disease. That is what happened as result of anti-fur efforts. Few sights are more pathetic than that of mud-bedraggled nutria—all skin and bones, and fur falling off in clumps—staggering to certain death across a mud-slick, pothole-dotted expanse of defoliated marsh, the result of kind but misdirected hearts.

The balance between trapping pressure and population increase has continued to seesaw. The activists lost some of their clout and fur began appearing on models again. In 1985, the harvest rose to just over a million from 700,000 in 1983. Trappers returned to the marsh after being laid off their jobs in the oil industry following the early '80s oil bust. The price rebounded to four dollars per pelt.

The return to trapping also provided an important lesson— that the marsh is a vital economic cushion as well as an economic resource. But then the oil patch recovered in the late 1980s and the cushion lost its usefulness. The harvest dwindled to less than 400,000.

Salvation appeared, at least for a while, from Moscow. A new market opened in the economic flurry following the fall of Communism. Russian designers and furriers discovered in nutria a cheap and plentiful buffer against brutal winters. Suddenly the price per pelt rose again, reaching $5.17 in 1997, the first season since 1980 that a good skin brought more than five bucks. Despite the continued good times in oil and gas, the number of pelts taken began inching up to almost 400,000 in 1997.

But good fortune was short-lived. Disaster returned, this one accounting for the present toll of 100,000 acres of damaged marsh. In September, 1998, just as trappers were beginning to think about fixing up their traps for the coming season, the Russian economy disintegrated in a pox of corruption, bad loans, and a default on its $13.5 billion debt. Orders vanished. Die-hard trappers, or desperate ones, harvested only 114,000 pelts that winter and received 50 cents each for them. The dealers who bought them could get rid of only half of them. Sixty thousand pelts are stacked up in various freezers awaiting buyers.

Noel shakes his head over prospects for the coming season (1999 to 2000). "There's no reason for anyone to trap nutria for fur anymore," he says. This worries him for a reason quite apart from the damage that will be wrought to the marsh. He sees, as do many people in tune with the rhythms of the marsh, the death of a way of life. It used to be that come November, when winds swept down from the plains, thousands of trappers and their families headed for the marshes, leaving their homes and pulling their children out of school for the winter. Out in the chill-killed marsh, they set up life in tarpaper-clad camps and spent the days running traplines and evenings skinning, scraping, drying, and stretching skins. This human migration seemed as embedded in the culture as the movement of the myriad waterfowl that blackened the marsh sky was instinctive. Both began and ended at the same time; both were driven, seemingly anyway, by forces that no one really understood.

Two decades ago, Clifford Stelly took his whole family out to Boston Canal leading into Vermilion Bay for the trapping season. I spent many days with the family throughout a winter, observing a tradition—now all but dead—being lived as though it were just part of life. The first time I met Clifford, he was standing on the rough-built dock in front of his camp. Corpses of nutria and muskrat surrounded him. He and three of his five sons had brought them in that morning from the traplines they had run on the marsh every fall and winter. He reached down a huge hand to grab my things and help me up from the boat. Then he stood, in a tilted-over fashion he always assumed, looking embarrassed in the midst of all the death. "This is a dirty business," he explained. "We skin-nin' now. We got a lot a nutr'a here."

His sons, Blaine, Wyndal, and Randall, each with a knife in hand, looked away from me shyly. They had set up a production line which, swifter than the eye could follow, separated pelt from animal. The pelts were flung into a bucket for soaking, later to be scraped, stretched, and hung in the drying shed. The naked corpses ended up on the dock where Charlyne, one of Clifford's three daughters, gutted them for sale to mink ranchers up north. The viscera ended up in the bayou, fodder for crabs.

As the three sons worked on the animals, Blaine offered that they could skin out fifty nutria in an hour. They left the muskrats, more difficult to skin, to Clifford. Sons tended to avoid the smaller muskrats. They were harder to skin than nutria, their pelts fine and easily torn. Clifford could skin out a muskrat in half a minute. Some older trappers could do three a minute.

I commented on the pace of the work. Wyndal looked up but then away from me. "Yeah, you jus' git used to it; that's all. I been doin' this ma whole life."

The cabin was a plywood affair up on stilts and surrounded by years of rusting traps, exhausted electrical generators, grimy oil drums, and undecipherable oddments—all contributing to the slight height of an island surrounded by marsh as far as you could see. Inside, Miss Della, Clifford's wife, was preparing a gumbo in a huge iron pot. Deena, their youngest daughter, and various daughters-in-law, followed Della's instructions, given in a singsong voice and carried out with a casualness that suggested a partnership rather than adherence to a line of authority. A vat of big crabs, their claws beautiful hues of blue and red, rested next to the gumbo pot, awaiting their fate. Shelly, the youngest child, about twelve, would charge in and out, sometimes carrying a freshly caught catfish, sometimes clutching his .22 rifle.

Eventually, the men filed in, hands washed as best they could. The steaming gumbo, crabs, now hot and red, the inevitable loaf of

Evangeline Maid bread, were waiting on the little kitchen table along with a couple of gallons of soda. Assembled, everyone kind of stood around, eyeing the food or glancing at the TV that never went off unless the car battery it was hooked up to died. Finally, Clifford would announce: "Dey got some food on the table over 'der." One by one, the family would fill their plates, Della always last. They ate in communal silence, except for brief comments about something that appeared on the TV. As plates were wiped clean, Clifford would give orders to his sons in ever so gentle a way, sometimes with just a word, Blaine to stretch some skins, Wyndal to check the catfish lines, Randall to repair some crab traps. The sons, so imbued with the routine over the years, just nodded in quiet acquiescence.

This movement of people doesn't happen anymore. "A whole cultural tradition is being lost," Noel says, before starting up the engine and heading across the depleted marsh. As we go along, he tries to avoid hitting nutria that, strangely, run toward the oncoming hull rather than away.

They might not be bright but they're good eating. At least people like Noel and a few Louisiana chefs are hoping to be persuasive on that point in an effort to find a new use for the ubiquitous animals. They have a tough battle. Louisianans look upon the prospect of eating nutria much the way a New Yorker would, say, pigeon. It's hard to eat something you see splattered all over the roads. It's hard to see an animal with a rat's tail as in any way appetizing.

Yet, the real glory of nutria is its meat, Noel preaches—22 percent protein, only 1 percent fat. Then he admits, begrudgingly, that "it's a big step for a housewife to buy nutria in a supermarket, I guess." Not big enough not to try. In desperation to find a market for the bothersome animal, the state Department of Wildlife and Fisheries has valiantly attempted to persuade both trappers and cooks to give nutria fricassee a try. Trappers are offered a dollar per nutria beyond the price they receive for the pelt if they sell to a facility licensed by the federal Department of Agriculture and state

Department of Health, which will then process the meat. In 1997, one processor sold 8,500 pounds, but not enough to dent the problem. The chef Paul Prudhomme got involved for a while, but bowed out when he understood the entrenched local attitude against the idea.

Wildlife and Fisheries officials are also flailing about the world, trying to do nutria deals to rid Louisiana of the animals. They've been talking up the Chinese. They flew to Japan to attend a food show outside Tokyo where Japanese purveyors were offered bits of nutria done up in different ways. It was not a success, Noel reports. The Japanese appeared interested but noncommittal.

Undaunted, he points out that back in 1972, when alligator hunting resumed after the reptile had begun to make a comeback, no one, but no one, ate its meat. Now, it's a fun food. You can buy it fried on a stick in the French Market and in po' boys in tourist restaurants, the kind that are filled with about a million stuffed alligators up on the walls that stare down at you as you eat their cousins. Noel hopes that nutria meat ends up with the same acclaim. He points out that in some eastern European countries, nutria has long been consumed.

The nutria is not moving at the mammoth Winn Dixie in Belle Chasse. Philippe Parola, a transplanted Parisian, frets. He is the latest chef commissioned by Wildlife and Fisheries to give nutria legs. A small man of great French sensuality, his eyes crinkle in a slightly inquiring smile, asking rather than looking. His wiry physique says as much in its subtle shifts as Calvin Klein hopes to impart in his underwear ads. He is definitely different from most people in the supermarket, who are rather larger. With Winn Dixie's blessing, and under contract to the Wildlife and Fisheries Department, he has set up a little nutria meat promotion center in a corner of the store. At the approach of shoppers, Parola picks up a paper plate of sliced nutria sausage and makes his offering in a Cajun-glossed French accent. To men he says things like "Saampul zee nutria, zee *ragondin*; you can help to save zee marsh." To women he is more deferential, delicately proffering the sausage-dotted paper plate with a graceful sweep of his arm and a little bow. "Mademoiselle, *s'il vous plaît*, theis ees the healthiest delicace known; pluease sample ete, you will be healing our enveerament. Thank you so very much." Most people chuckle and move on, seeming not to know how to respond to this anomaly among the towering displays of soft drinks and snacks.

Parola spends two or three afternoons a week in different super-markets across the state, giving away nutria samples and hopefully selling vacuum-sealed packages of nutria sausage at $3.99 a pound. He says he sold 20,000 pounds in 1998. The rest of the time, he runs his cooking school, Louisiana Culinary Institute, and his restaurant, Bear Corners, in Jackson, Louisiana, where he offers nutria on the menu among more standard fare. He lists it by its more attractive, *très français*, French name—*ragondin*—as in *ragondin à l'orange, culotte de ragondin à la moutarde*, or *soupe au ragondin*. I have not had the plea-sure to sample Parola's dishes, but I have eaten nutria prepared as one would *coq au vin*. This was *ragondin au vin* and it was delicious; the meat slipped off the bones in moist chunks that imparted, in my mind, anyway, just a suggestion of sweet marsh grasses.

Parola describes his present circumstances as not quite as attrac-tive as when he arrived in this country in 1981. He tells me he came here at the behest of relatives of Carlos Marcello, the celebrated Louisiana Mafia godfather who is reputed by some to have arranged for John F. Kennedy's assassination. Only after Parola arrived did he learn the identity of his boss. Among Marcello's sprawling empire, which included casinos, banks, liquor stores, shrimp boats, and motels, he owned four restaurants in New Orleans, some of them of high repute, like La Louisiane, and Broussard and L'Enfant. He needed a chef, a French chef. Parola, who was working restaurants in Italy, was tapped by some of Marcello's cousins to come to America and run the restaurants. He came with the French flair for commin-gling flavors which he adapted to the spiciness of creole and Cajun cooking. The restaurants thrived and Parola became known as one of the young, innovative chefs of New Orleans. But the glory was short-lived. Even while Parola was building up the restaurants, federal prosecutors were closing in on Marcello. For three decades, he had managed to escape prosecution by buying off juries. In 1983, he was finally convicted of racketeering and sent to prison. His empire began to collapse, his restaurants included. When he died in 1993, the restaurants closed and Parola found himself on his own.

Selling nutria is more challenging than serving finely arranged entrees. A woman grabs a slice from his paper plate and pops it in her mouth before Parola has a chance to tell her what it is. When he does, she gags and spits it into a tissue. Parola comforts her and informs her of her "erreer"—and he tells her that this nutritious bounty of the marsh is for her enjoyment. Some shoppers down a slice and walk away shaking their heads as though they can't believe they have just eaten such a disparaged animal. Others recount the numbers of squashed nutria they encounter on roads every day and tell Parola that they are not about to eat roadkill. Some just say, "ugh," and huff off. It's hard work, selling nutria in Louisiana. He should go to New York or a place where people don't know what a nutria is.

He tires as the hours go by, his sharp eyes lose a bit of playfulness. He takes to the aisles one last time with his paper plate held aloft, looking as if he must be serving the most exquisite delicacies to royal diners. Ten minutes later, he returns to his promotion center with the plate's contents mostly untouched, his chef's hat askew.

As I leave, better educated, but not hopeful that the American palate is going to be swayed toward nutria anytime soon, Parola perks up and says wistfully: "Remember, *s'il vous plaît*, that ete was a French chef who solved zee problem with La Louisiane's enverrament."

Not quite. A couple of months later I am flying down Calumet Cut, a navigation canal near Morgan City, with Peanut in his work boat. This time, we are going nutria-hunting, now that Peanut has found someone who will buy furs. After searching the entire state for a buyer, he located a fur dealer in Alexandria willing to gamble that the market will come back. The buyer agreed to take six-hundred skins at two dollars each, hardly worth the fuel to get to them.

But then, Peanut is one of those dwindling numbers of people who just can't stay out of the marsh. Anyway, he can scrounge a living from it, he takes.

While it was a triumph of sorts that he has a buyer, he has another problem of far greater magnitude—the weather, or maybe the climate itself. Though the winter temperature in South Louisiana rarely falls below forty degrees, humidity is always in full force, resulting in a penetrating chill that cramps and stiffens fingers, arms, and legs. The chill puts nutria on the move, particularly at night, scurrying along well-defined marsh passages, doing what they do best—eating. Where the "roads," as trappers call them, enter a bayou, they set steel traps for these largely unwary animals. Early each morning—before the collapse of fur trapping—trappers hauled themselves, stiff-jointed, out of bed and away from their ragtag marsh camps to run their lines. The bayous were often ice-encrusted and the marsh grass looked as if it had been dipped in fluorescent white against the rising sun. Sometimes they'd find nutria in the traps without tails, frozen off during the night when ice formed around them.

During some winters, portions of the marsh froze over entirely, preventing trappers from traveling in their mud boats along *trainasses*, wide ditches that allowed them access to their traplines. That happened a few times when I used to stay with the Stelly family, fourteen or so people in the little plywood cabin. After a day or so of waiting for the weather to lift, they'd fight the piercing cold and run their lines on foot, returning hours later hypothermic and shaking uncontrollably.

Now, two decades later, the winter weather is very different. In fact, the winter of 2000 is eerily warm. Already, in mid-February, the marsh is greening, two months ahead of schedule. The 70-degree midday temperature slows the nutria down; their hormones signal them that the warmth brings a ready harvest of fresh roots and shoots right under their feet. No foraging necessary. Traps are not effective under such conditions when nutria are not on the move.

But there's something else that deters Peanut and alters the pull of winter tradition. It's cottonmouths and rattlers. Warm days bring out more than new vegetation. "I wouldn't wanna walk in that marsh for nothin' today," says Peanut emphatically. "That grass is fulla cottonmouths. They're dangerous, they are."

So today, Peanut is adopting a new harvesting technique. He has brought his .22 with him and aims to shoot a mess of nutria. After about half an hour of winding down bayous, coots rising at our approach, thick as black flies from dead meat, we arrive at his hunting grounds—the 140,000 acres of the Atchafalaya Basin Wildlife Management Area—which he leases from the state for trapping, the same area where he hunts alligators.

At the edge of the Gulf, it's a desolate and wildly beautiful expanse of ragged marsh still winter-dead, brown and full of skeletal reminders of last summer's luxuriant growth. It smells of mild, hopeful rot. In the near distance, a strange cacophony is pervasive, inescapable, and lovely, and a million geese arise from a feeding ground, filling the horizon with sheets of darkness waving against the gray sky. They're blues, and snows, and speckled bellies and they won't stop rising until they fill the sky leading to the Gulf, smudging the horizon the gritty hue of kerosene smoke.

Soon, Peanut noses his boat into narrow twisting bayous. We are close enough to the banks to see nutria hunkered down beneath bushes and grass clumps. Some stay put at our approach; some waddle to the water and swim across the bayou, an escape tack of utmost folly, for it puts them within a dozen feet of the boat. Peanut is not a very good shot, really, for all the years he has spent in the marsh. He takes aim and fires at one swimming ahead of the boat. The bullet smacks the water a couple of feet away from the animal's head. The second and third bullets raise little geysers. The fourth one hits, but amazingly, the bullet ricochets off with a whine, causing the animal at last to dive. It's going to surface somewhere, but where, we don't know, so we spend the next two minutes playing a guessing game. Finally, it's up against the far bank and Peanut dispatches it into a quivering mass of nerves with a final, and mercifully accurate, shot to the brain.

He hauls the twenty-pound corpse aboard and runs his hand

along its belly, happy in the obvious luxuriance of the silken fur under its guard hairs. "These here are good nutria right here," Peanut exclaims. "It's a shame you can't get decent money for 'em anymore."

In the past, Peanut has killed two hundred animals a day in his traplines. Collecting this number of nutria is an all-day job from dawn to dusk. Then, he had to skin them out or sometimes he skinned them as he hauled them into his boat, laying out the pelts in an ice chest and flinging the carcass into the bayou to be fed upon by coons and crabs. Most often, he did this with friends, and as they skinned they would tell stories and exchange marsh lore. That's gone, now that the winters have warmed and the market has died.

Peanut's limit of six hundred nutria is far too minor to even dent the population. I see evidence of this after I have left Peanut and am driving back to New Orleans. My eyes are drawn to a soggy field next to the road. Little shapes dot it; nutria by the hundreds are in the process of destroying it. The previous night, some apparently decided to cross Route 90 to sample a field on the other side. They never made it. Nutria pancakes litter the asphalt, but not in sufficient numbers, of course, to stop the invasion. And the sight certainly is not an inspiration toward gustatory delight.

Chapter Eight

The Mistake

Early morning, May 3, 1978. Something is wrong. The pecan tree in the yard is flaccid, its customary day-greeting flounce of leaves, in the rising breeze, stilled. The walls and ceiling of the apartment in New Orleans I then rented are heavily dank; everywhere, moisture clings. No sounds of traffic on St. Charles Avenue either, at 8 A.M. usually bumper to bumper, trapped faces herding toward New Orleans's Central Business District and French Quarter for another day in offices, behind counters, smiling to tourists. None of the St. Charles streetcar's delicious lurching and grinding noises. Water—tripping over the old clapboards, gutters long since rotted out or collapsed—streaming down windowpanes onto rotted sills, gurgling through widening crevices in the building's dilapidated roof—the sounds of water are invasive.

Blurs of shapes suggest themselves through the rain—of live oaks, the colonnades homes across St. Charles, of paralyzed traffic, a stranded streetcar, all obscured in hazy aqueous blue. But I can see water on the pavement, a black flow with body and depth. It's moving; there's current pushing it. At the cross streets—Jackson, Philip, First—water surges over the streetcar tracks, creating miniature plumes and horsetails. Soggy human forms wait, drenched on the median strip, for a streetcar that cannot carry them for the flooding, umbrellas collapsed under the deluge.

I, a visitor to the city and state for the purpose of writing about it (my first encounter with this soggy environment) am excited. I have no stake in what is happening. My only duty is to observe, amazed that a city can fill up so fast with water.

Over 10½ inches of rain fell that day, released from a battalion of smoldering fronts sweeping out of North Texas. By midafternoon, canoes and outboards were trafficking St. Charles. Water covered the floor of my car, parked on higher ground away from the subsided pavement. Vehicles on the avenue were derelict, abandoned in angles of disarray, their engines drowned. Life stopped that day in a good part of New Orleans. Uptown was a mess; Mid-City verged on returning to the marshy state from which it emerged in 1718. The French Quarter's narrow streets now flooded, it was easy to see why their blocks in eighteenth-century city plans were called islands, and the paths beside them, banks, or in French, *banquettes*, the name given to sidewalks in New Orleans into the 1960s.

A million or so calls were placed to the Sewerage and Water Board, inquiring if the pumps were working. As many assurances were given that they were, that the floodwaters were being pumped out as fast as possible. What could people do but wait it out? New Orleanians, as is their wont, made a party of it, affirming once again the Big Easy's familiar appellation. They splashed around in the streets, cruised them on air mattresses, drank some beer, played some tunes, and had a great time waiting for the waters to recede.

By the following morning, St. Charles was open to traffic as usual. But residents of some low-lying neighborhoods were in water for three or four days. Snake warnings abounded. Cottonmouths were the ones to fear. They like to seek shelter in dry places, like shelves and upstairs bedroom closets. There was the usual worry that alligators would invade playgrounds. The anxiety was groundless. Alligators are never in a mood to wander around strange neighborhoods, biting on toes, particularly when upset by a sudden change in their habitat. People wondered if the remains of dead friends and relatives had suffered, ensconced in their flooded, above-ground vaults in the city's cemeteries. Even if the waters had wet their bones, they would certainly be better off than if they had been interred beneath the soggy soil.

Flooding in New Orleans is not uncommon, though this was one of the worst on record over the past several decades. But it was not up to the May 1995 flood when over seventeen inches fell on some parts of the city. And then in September 1998, tropical storm Frances came along and dumped almost 32 ½ inches in a three-day period.[1]

It is not a particular point of pride to New Orleanians that their city is situated in the worst possible location. It is just a result of fate and history. The amplitude and power of the Mississippi River that borders the city's south side, and the bulk of Lake Pontchartrain to the north, hold in their grasp the easy threat of devastating the city in a matter of hours. Sixty percent of New Orleans lies below sea level, as much as eight feet in some places, particularly along the four-mile stretch of North and South Broad Avenues, and digging itself in deeper as sea level rises and the land subsides.

There are famous examples of subsidence's quirky doings, particularly in the subdivisions of East New Orleans—of homes built on concrete slabs in the 1970s, now beached and awkwardly balanced as the land around them sinks. Front steps have lost their relationship to front doors and asphalt driveways ram into the edges of concrete carports. Evidence of the compaction of the muck upon which the city was built is everywhere, imparting a look of delectable decadence and the nice pungent odor of wood rot. Wayward vegetation pokes through collapsed sidewalks; massive live oaks lean with gentlemanly drunkenness over round-surfaced streets with gutters at their edges sunken into absurdity; walls of older structures slump out of plumb; tomb vaults split open in the

---

[1] New Orleans receives an average annual rainfall of fifty-eight inches, making it the wettest city in the country. Storms tend to move rapidly and locally over the metropolitan area, not unusual in itself. One part of the city may thus receive a torrential downpour, whereas an area two miles distant may get no rain. Because drainage depends on pumps, rather than gravity, those areas that do receive substantial rain in a short period tend to flood easily.

city's cemeteries, spilling bones that jut up from the sinking land.

Another result: the inexorable thrusts and pressures of subsidence have squashed storm drains and the myriad connecting pipes and tunnels into partial ineffectuality against the rampages of storm runoff. But this makes no difference in the aftermath of large floods, the kind that could accompany "The Big One," as people in South Louisiana have come to respectfully refer to the devastation that is inevitably their fate. It is fair to say that given the city's siting and its particular topography, no drainage plan, no matter how exquisite its engineering, could deal with the volume of water that could deluge the city under the power of a storm that, like the perfect boxing punch, hits with the right speed, from the right angle, and with extraordinary power.

How New Orleans's situation has come to be requires a dip into the city's past. At the location of the present French Quarter, the Mississippi throws itself into a tortuous twist. The current here wrenches itself from a northerly direction and shoots itself directly east. Spring runoffs over the past one thousand years or so, in which the river has occupied its present bed, have sent untold tons of mud-laden water overflowing the bank here. By the eighteenth century, a firm and fertile levee had accumulated, reaching a height of fourteen feet, a workable platform for settlement. Even better for early settlers, the levee was wide, angling gently half a mile or so away from the riverbank. Nevertheless, the city flooded periodically, compelling its nervous citizenry to add repeatedly to the levee, year by year. The levee—or concrete floodwalls abutting the river—stands today twenty-three feet above sea level.

While the Mississippi's floodwaters have always threatened the city from the south (the south, due to the river's serpentine meanders so that at New Orleans, the river actually forms the city's southern, rather than western, edge), the greater danger is to the north. There lies the expanse of shallow Lake Pontchartrain, a roundish, flat-bottomed pan of water forty miles across at its widest point. Floodwaters are not the main threat from this direction; giant waves are, pushed by walls of wind, ghastly to contemplate.

• • •

Two and a half centuries ago, the slope of the Mississippi's natural levee, descending away from its banks, bottomed out in a marsh where North and South Broad Avenues traverse the city today, 1½ miles from the river at the closest point. Drainage of this once soggy area was compromised by the small levees of a former distributary of the Mississippi that branched off the river near today's Causeway Boulevard, meandering southward to empty into Lake Borgne.

By the time the French arrived, this distributary channel had dried up, leaving two parallel levee remnants. The view from atop one of the giant cypress trees, which used to forest the area between the river and the lake, might have suggested the work of two giant moles burrowing side by side. Today, the remains of these levees are known, if they are known at all, as the Metairie-Gentilly Ridge. They provide the bed for Metairie Road, City Park Avenue, and Gentilly Boulevard.

North of this ridge, the 2½-mile stretch to the lakefront angles downward to six feet below sea level before rising sharply to a slight levee bordering the lake. Like most levees in South Louisiana, the Pontchartrain levee has been heightened over the decades; today, it is eighteen feet above sea level and being added to in preparation for The Big One.

Between these sources of potential deluge—the Mississippi on one side and Lake Pontchartrain on another—lies the city, waiting and fearful, an apparently willing accomplice to its own drowning due to its accidental siting so long ago. The Sewerage and Water Board of New Orleans—or S & WB, in daily talk—the entity charged with getting storm water up and over the levees, doesn't mince words in describing the city's plight. Its handouts acknowledge the misfortune of the city's location at the bottom of a "saucer." A little cartoon has a colonial-day, costumed figure sitting in a bowl. He is supposed to represent New Orleans and he wears a puzzled expression. His cartoon balloon says: "In 1718, Bienville founded me here . . . in a bowl surrounded by water." An Indian in a canoe off to the side quips: "Even us Indians wouldn't live here." The message carries a blatant visual effect.

The question becomes: despite the higher ground created by the natural levee, why did the French ever found New Orleans in this particular miasma? It may well have been an accident—of lethargy, shortsighted convenience, of the need for human contact in this eerie land in the early eighteenth century, which for Europeans must have been a nightmare of rash-spreading, foot-rotting humidity. But the presence of Bayou St. John was certainly a decisive factor. It provided a nearly complete link to the Mississippi from the Gulf via Lake Pontchartrain, knocking off 110 miles of struggle up the Mississippi against surging current, shifting sandbars and a barrage of sunken logs and uprooted trees. But there were and are other places upriver that make as much sense.

Today, Bayou St. John commences abruptly near Mercy Baptist Medical Center, though in early colonial times, its drainage basin included rivulets and lesser bayous that snaked through what is now Mid City and Uptown. By the time the now-truncated bayou reaches City Park on a gracefully curving course to Lake Pontchartrain, it has widened to forty feet and in some spots appears so undisturbed, you might think New Orleans never existed.

Toward the end of the seventeenth century, the French and Spanish fiercely competed for the Gulf coast. Steadily might be a better word than fierce, given slow boats, ponderous weapons, and kings who did not make quick decisions on the merits of colonizing swamps. In 1682, René-Robert Cavalier, Sieur de la Salle, a French Jesuit-trained math professor turned Canadian explorer and fur dealer, ran the river to its mouth, claiming everything he saw for Louis XIV and naming much of the countryside *La Louisiane*, in honor of the king. The voyage's purpose, perhaps boosted by notions of conquest, was principally fueled by La Salle's ambition of setting up a fur-trading empire along the river.

La Salle organized another expedition, this one ostensibly to put up a settlement at the mouth of the Mississippi. He sold the

French king on the idea by saying that a colony could serve as a military base for raids on Spain's gold and silver mines in Mexico. He forgot to mention distances, though, leaving French ministers with the impression that Mexico was a mere stone's throw away.[2]

The voyage, blessed and financed in 1684, was an utter disaster. All four ships missed their mark, an impossibility today, but testament to the navigational crudities of that same era that took the pilgrims to Plymouth, rather than further south to a destination named in the Mayflower Compact as "Ye Northerne parts of Virginia." The wayward French ships bypassed the Mississippi's mouth and landed in what is now Matagorda Bay in Texas.

La Salle must have quickly realized that the Colorado River, which empties there, was not the Mississippi. But the expedition was beyond salvage. Three of the ships were virtual wrecks; the fourth headed back to France. La Salle set out on foot for the Mississippi with some men. They wandered the coast for two years, up to their armpits in muck, mosquitoes, and imagined monsters. It must have been a trip out of hell. In 1687, La Salle decided to give up and return home to Canada, on foot. The decision was fateful. Stretched beyond their limits, morale sinking with every step, his remaining men mutinied and murdered him.

Even after death, La Salle's efforts to sell the idea of colonizing the lower Mississippi lived on. The French government, running out of money, was still under the impression that the Mexican mines were next door. A colony was necessary both as a military base from which to loot the mines and to discourage Spanish encroachment on the Mississippi. In 1698, Louis XIV commissioned Pierre le Moyne, Sieur d'Iberville, to head an expedition to the mouth of the Mississippi. A distinguished French-Canadian naval officer who had made a successful career harassing the British in North America, his mission was to legitimize La Salle's prior claim to the Mississippi River valley by establishing a minor military presence at the river's

[2] Carl A. Brasseaux's *A Comparative View of French Louisiana, 1699 and 1762* provides an excellent view of early exploration along the lower Mississippi.

mouth. Along with his younger brother, Jean-Baptiste de Bienville, Iberville ascended the river in 1699 as far as the mouth of the Red River, some one hundred miles northwest of the present New Orleans as the crow flies, but over twice that distance as the river flows. As the banks grew higher, he must have spotted some secure sites for an outpost. But he returned downstream without planting the French flag in the rich soil.

The ascent, by longboat and canoe, must have been as fascinating as it was frightening. Iberville kept a daily journal; his entries echo the loneliness of the lower Mississippi in the early seventeenth century, "shrouded in fog" and "teeming with driftwood." While most of the Indians encountered along the way were curious about the Europeans, their ephemeral nature must have been disconcerting. They watched in groups from the banks; the next moment they had vanished. "At midday, we discovered a campfire on the left bank, in the prairie," he wrote, "and we know that an Indian had been there. . . . My brother, with two canoes, always sails along one of the river's banks and keeps watch for traces of the Indians."

Despite the Indians' elusiveness, Iberville managed to persuade an Annochy Indian, a branch of the Sioux, to guide the expedition. The man was given a hatchet in exchange. It was this unnamed Indian who turned a casual observation into a chapter of U.S. history. He "revealed a terminus of the portage from the southern shore of the bay [Lake Pontchartrain]," Iberville writes, "where the Indian boats land in order to descend to the river. They drag their canoes along a fine path, where we found the baggage of people who are either leaving or returning by way of this portage. This Indian, our guide, took a parcel there. He remarked that the distance between one end of the trail and the other is indeed inconsiderable."

Iberville was probably referring to Bayou St. John, although he gave it a location that was farther upriver from where New Orleans was to rise. This discrepancy, pointed out by Carl A. Brasseaux of the Center for Louisiana Studies at Louisiana State University at Lafayette, suggests that the bayou shown to Iberville might have been Bayou Trepagnier, which looks today very much as it must have looked in 1699. At least, on the surface it appears unchanged, bounded on both sides by forests of green ash and its levees thick with Roseau cane, impenetrable reeds that cloak the edges of the Gulf's wetlands. Beneath the bayou's serpentine twists, from Lake Pontchartrain to within a half mile of the river at the present-day town of Norco, however, its bottom is horrendously fouled by an eight-foot layer of toxins from the Motiva and Shell Chemical refineries in Norco, twenty miles upriver from New Orleans. Shell, which now operates the Motiva refinery in partnership with Texaco and Aramco, permitted its refinery wastes to flow into Bayou Trepagnier for decades, a charge which it readily acknowledges. While environmentalists clamor for a cleanup, Shell claims that the toxins are dissipating on their own. It's hard to appreciate this, though; stick a canoe paddle into the bottom, churn it around, and horrible black stuff oozes up. The nearby air suddenly turns thick with the smell of petroleum.

The discovery of the three-mile portage and bayou, whether Bayou St. John or Bayou Trepagnier, led Iberville to reconsider the merits of defending the lower Mississippi. No one in his right mind would choose to slog up the river when he could traverse the far more peaceful lake, enter a bayou, and carryover to the river with the added benefit of finding himself over one hundred miles upstream of its mouth. Iberville must have realized that traders or invaders would use the lake to get to the river, and it was that route that had to be fortified against the Spanish, who would be approaching from the west, probably from Pensacola. The forts that Iberville built in what are today Biloxi and Mobile guarded the probable route the Spanish would have used to gain the lake.

Explorations of the lower Mississippi accomplished, Iberville returned to France several times to look for funding for a settlement. Brother Bienville always remained in Louisiana. What he did for the next eighteen years is something of a mystery. He is thought to have clung to the vicinity of Bayou St. John. French traders and trappers passed through on occasion and a Houma

Indian village was supposedly located at its headwaters, marked today by Encampment Street.[3] His brother had little success founding a settlement on the Mississippi, and died without fulfilling his mission.

Not until 1716 was any real progress made on the settlement. In that year John Law, the maverick Scottish financier, persuaded the Duke of Orleans in France to give him free rein in reviving the French economy, weakened and debt-burdened by years of religious wars, food shortages, and the extravagance that went into building Louis XIV's palace at Versailles. Law formed the Mississippi Company to which the government gave a monopoly on trade between France and Louisiana. All this from a gambler who had fled England after killing someone in a duel. He must have been a hell of a marketing man. His spiel was one of the glories of the New World; he talked up what he called mountains of gold in Louisiana and a huge emerald in the Arkansas River. The response was a stock grab, running shares up to the sky before the Company had brought in a *sou* of the promised riches.

Law's marketing attracted hundreds of colonists to the Mississippi valley. They came to an uncertain life with no guarantees. In a foreshadowing of what was to be repeated a number of times during the next three and a half centuries, the river flooded in 1718 and destroyed the nascent settlement of New Orleans, what there was of it—a few thatched huts and gardens. The flood

---

[3] The name Houma, like the names of many Indian groups, is an Anglicized rendition of what the Indians were *thought* to have called themselves. Actually, La Salle referred to the Houmas as Chouchoumas, a probable misunderstanding of *chakchiuma*, their word for crawfish. When La Salle descended the Mississippi in 1682, this Indian group hunted in the area around present-day Baton Rouge. The Houmas separated their territory from that of the Bayougoulas, downstream, with a red stick (*baton rouge*) stuck in he ground, or so the story goes. At some point, they lost this area to the Taensas. Some then moved to the Bayou St. John area. When New Orleans began to be established, they went further south into the marsh around present-day Houma, which is how the city got its name. Though their descendants intermarried with the melange of races and nationalities that settled in this region, the Houmas are the largest Indian group in Louisiana and are seeking federal recognition.

might have called into question, however, the wisdom of locating a city where it has grown today. An itinerant architect named Le Page du Pratz, one of the few newcomers to keep a written record, noted that "serious thought was given to relocating New Orleans upstream to higher ground." Nothing came of it, however, especially as three hundred settlers arrived later in the year, land grants from the crown in hand, wealth gleaming in their eyes, and began clearing land around Bayou St. John.

By 1720, investors realized that puffery was the only reality behind Law's promises. As has happened countless times since, and is sure to happen countless times again, investors panicked. They sent the stock tumbling, pushing themselves into poverty with the exception of the few with exquisite timing to sell just before the crash. Law fled again, this time to Italy, where he spent the rest of his life as a gambler, dying destitute in Venice.

His bubble burst, but the colonization it fomented began to swell. By early 1722, New Orleans consisted of, so a Father Charlevoix reports in a letter, "a hundred huts of wood, two or three houses which would not embellish a village in France, and half a wretched warehouse that they [settlers] had consented to assign to the Lord." Not for long. In September 1722, the settlement was again storm whipped and wiped out, to the evident delight of John Law's designer, Adrian de Pauger. De Pauger had drawn up a plan for the city the previous year, but no one had paid any attention to it, erecting structures wherever they pleased. The storm cleansed the hodgepodge, allowing a rebuilding that largely followed the design of the French Quarter as it appears today.

The natural levees along the Mississippi were heightened, decade after decade, by landowners whose property faced the river. These levees protecting New Orleans from the Mississippi have always held. Flooding, however, has always been as much a part of the city's life as Mardi Gras. Between 1735 and 1927, the year of the highest water ever seen on the lower Mississippi, the city flooded nine times, due to levee breaks—or crevasses—upstream.

Then in September 1947, a hurricane passed right over the city and hurled up a storm surge that leaped over the Lake Pontchartrain levee, dumping 2 ½ feet of water into the city and killing thirty-four people. It happened again in 1965. Hurricane Betsy, tracing one of the most contrary courses of any hurricane to strike North America, had executed a complex series of about-faces four hundred miles north of Puerto Rico before straightening herself out and churning northwestward toward the Carolinas. Then she performed another dazzling acrobatic, baffling meteorologists and raising adrenaline levels along the entire eastern seaboard. After completing a second loop-the-loop, she set off south for the Bahamas but then turned sharply west, north of Nassau, brushed the tip of Florida, just avoiding Key West, and veered west by northwest straight for New Orleans. But she spared the city, just barely. As darkness fell on September 9, she swept over Grand Isle, drowning the island and destroying almost all its buildings. She roared up Barataria Bay, pushing winds of 150 miles per hour, grazing New Orleans to the east and Thiboudeaux to the west, flooding a half million acres, killing eighty-one people and wreaking $1.4 billion in damages. She then headed north on an unprecedented inland journey, which took her all the way to Little Rock, Arkansas, before she was downgraded to a tropical storm and collapsed just short of Pittsburgh.

The hurricane dropped a paltry four inches of rain on New Orleans, but gave the city its worst flooding ever. Her most powerful weapon was Lake Pontchartrain's waters. Easterly winds built them up into massive battering rams. The levees protecting the north side of the city became mere inconveniences to surges from the lake which poured water into East New Orleans, filling it to house-rooftop level from the Inner Harbor Navigation Canal to the shore of Lake Borgne and drowning over forty people.

Betsy brought New Orleans face-to-face with the uncomfortable reality that its location might be a big mistake. But what can one do at this point? East New Orleans remained a morass of soaked carpets, derelict cars, and floating driftwood long after the storm had receded. Days after small towns and cities, places like Lafitte, Barataria, and Paradis, had righted themselves, drained themselves, and picked themselves up, parts of New Orleans were still afloat or buried under feet of murky water.

The Sewerage and Water Board was frantic before, during, and after the storm. The frustration coursing through its pumping stations as the waters rose, recorded as a postmortem by a D. D. Modianos, the board's mechanical engineer, not only speaks to the hours during and after Betsy, but serves as a warning of what could happen in the future, despite improvements in both pumping capacity and communications. When levees give, or when storm surges burst over existing levees, the fact that nature has been in control all along becomes obvious horribly fast. The report reads:

> We were certain now that there were levee breaks on the Industrial Canal (called also the Inner Harbor Navigation Canal), but the exact location and nature of the breaks were unknown. Communication with the main office by phone and radio was fragmentary, but what communication was available indicated that none of the responsible agencies, such as the Levee Board and Civil Defense, could provide us with a coherent picture of what happened. We therefore requested the main office to obtain for us an army "duck" so we might reconnoiter.
> ... What was known, however, was that the situation on the east of the Industrial Canal was hopeless as all pumpage had been lost, whereas on the west side Stations 3 and 4 were pumping as much as possible from the area and were maintaining the water at their suction basins with reasonable margin to the station floors. ...
> ... Meanwhile, the surging flood waters inundated the Central Yard at Peoples and Florida Ave. to a depth of six feet or more in the streets, trapping a large force of S & W. B. personnel that had been assigned to the yard for emergency duty during the hurricane. More than 125 automobiles belonging to these men were completely submerged in the yard's parking lot.

That was in 1965. The threat of equal or worse inundation resulting from hurricane winds pushing water into the city has not lessened. This is not to say that the S & WB is not doing an impressive job. It is. But its function is to clear the city of the fifty-eight-plus inches of rain that falls on it each year. Hurricanes are not necessarily within its province. When you imagine a city that is mostly below sea level, that receives almost five feet of rain per year, and that drainage experts describe as a saucer, you have to wonder why there's any place to walk on dry land in the city of New Orleans. At least I did, and in reply, the S & WB provided me with a sheet of impressive facts including:

- The city's twenty-two pumping stations, all located in the lowest areas, are capable of pumping 35 billion gallons of water per day, mostly into Lake Pontchartrain. This is comparable to the volume of water in a ten-foot-deep lake with an area of eight square miles.

- Another way of looking at it: the system can pump three inches in five hours—one inch the first hour, and as the ground becomes saturated, one-half inch per hour for the next four hours.

- Still another view: the system is capable of pumping more water in twenty-four hours than is in the Thames River as it passes London.

- And still another: the combined capacity of all the pumps could fill the Superdome to brimming in the space of thirty-five minutes.

- A 185-mile-long network of covered and open canals throughout the city channels storm runoff to pumping stations. Some of the canals under median strips of the city's major thoroughfares are the width of two buses. A new canal excavated under Napoleon is wide enough to fit three buses side by side.

- The S & WB has its own electrical generating system with which it operates its pumps. Oddly, the electricity generated is at twenty-five cycles, following the specs of the original pumps, still in use. Commercial electricity is generated at sixty cycles. If, however, the board's generators are knocked out, electricity can be tapped from Entergy, the commercial provider, and converted on site to twenty-five cycles.

- The S & WB also has its own weather radar for spotting the approach of rainstorms.

Impressive, but enough? I was subsequently directed to the Meloponome Street Pumping Station (now called Drainage Pumping Station No. 1) at the intersection of South Broad Avenue and Martin Luther King, Jr. Boulevard. This gloriously hued 1850s brick structure, in one of New Orleans's lowest sections, drains some three-quarters of the city's uptown area. Transported by gravity, rainwater courses down from the Mississippi's levees, from the Garden District, from Audubon Park and the Tulane campus and through the thoroughly built-over lowlands that Bayou St. John formerly drained. These millions of gallons of water gather in total darkness in a completely hidden reservoir, two blocks in area, beneath South Broad at the base of the pumping station.

Inside the cavernous station, the pumps are immense and adored. Aficionados of drainage systems (it turns out there are many of them) marvel at their steadfastness, their look, their hugeness, their history. Visitors come in clusters from around the world to stare. Designed and built in 1913 by a New Orleans native and former S & WB superintendent, Albert Baldwin Wood, no pump has ever equaled the efficiency of a Wood screw pump. The original pumps remain the city's lifeline to dry ground, though their numbers have been added to over the years, still employing Wood's design.

New Orleans a century and a half ago was a fetid, stinking place with stagnant water simmering in clogged and haphazardly located drainage canals. Mosquitoes swarmed everywhere. Yellow fever was a steadily rising agent of death, killing at least 100,000 people in the 1878 epidemic. The cholera toll was devastating. The city, remarkably lethargic in dealing not only with disease but also with filth, had installed only four paddlewheel pumps by the end of the nineteenth century. Inefficient as they were, they at least got the water moving and even lifted some of it out of the city and into adjacent wetlands.

Not until 1899 did this enormous problem begin to be rectified with the formation of the S & WB, which was granted autonomy from the city administration. Its mandate was to drain the city. And it was given the privilege of getting down to this business without having to endure the usual processes of public hearings and political maneuverings. In that same year, Baldwin Wood graduated from Tulane with an engineering degree and found a job as an electrical inspector with the S & WB. He worked there for the next fifty-seven years. Wood's main passion was moving water from one place to another. His only slightly less consuming passion was *Nydia*, a gaff-rigged sloop, some thirty-five feet on the waterline, to which he retreated most weekends for long solitary sails from his weekend home in Biloxi, Mississippi. Between his pumps and sloop, New Orleanians came to love this unassuming man who merely went about his job and sought no glory.

He bequeathed his *Nydia* to Tulane. His will reads: "It is hereby an obligation and provision of this will that my boat the *Nydia* and her spars shall be carefully preserved . . . by Tulane University for at least 99 years." I know this because the bequest, inscribed on a plaque, rests against the *Nydia*'s cabin in her glass-walled sarcophagus tucked away in a niche of the campus and almost concealed by a stand of vivacious ginger bushes.

She's beautiful, naked in her cradle set in the middle of her tomb, surrounded by air, an ethereal substitute for water. She must have sat sleek and graceful, low on the water, sporting a downward curving bowsprit and bronze-trimmed portholes. She hasn't been cared for much since her owner's death in 1956. Her paint is beginning to loosen on her hull, revealing the outline of her planking. Water from leaks has streaked the floor of her tomb and dried, leaving the outlines of dirty puddles. Splotches spoil her smooth decks, layered in dust. Something has disturbed the dust on the stern deck. Footprints—as if someone had alighted on the boat and fled. Wood's presence, as modest in death as in life, can still be seen and felt by those who care to look for it.

As a young S & WB employee, Wood's duties were to inspect the pumps that the board had just installed and to keep them running. They were vertical pumps—Archimedes screws—which lifted water from low-lying drainage canals by means of an eight-foot-diameter, fast-rotating corkscrew to higher canals, and sent it to Lake Pontchartrain. Better than paddlewheels, they were still inefficient clunkers that couldn't possibly keep up with the water. Wood took it him upon himself to work on a novel pump design in his spare time. It was to be a horizontal pump that sucked and lifted water through an impeller—like a propeller in looks except that it pulls rather than pushes water. Its efficiency came in the powerful vacuum created, giving water no choice but to flow uphill into the vacuum and into the impeller, which spewed it out the other end into a higher canal. Perhaps in desperation, the board in 1913 agreed to the design, and the city put up $160,000 to have thirteen huge pumps built, each with a twelve-foot-diameter impeller.

In 1915, they were installed in the Meloponome Pumping Station. They are still there, and the people of New Orleans still depend upon them. Wood was not offered a penny for his extra work, and apparently didn't ask for anything. In fact, he gave the board the rights to the design in perpetuity. Such generosity has helped ensure Wood a high place in the collective appreciation of New Orleanians and pumping mavens everywhere.

From the gloom of the Meloponome Pumping Station early one morning, Joe Puglia emerged to greet me. He was dressed in a blue blazer and khaki pants, perhaps the requisite garb for a public relations official representing the S & WB, but out of place against the brute size and raw mechanics of the pumps. Heavy, indestructible iron pipes arch up from the invisible reservoir outside and below the building, pass through the cavernous shadows, and angle downward into the Washington-Palmetto Canal, which eventually empties into Lake Pontchartrain.

One of the first things Joe said, patting the shoulder of one of these behemoths, was: "You can't even get a piece of paper between the impeller blades and the inside of the pipes; that's how efficient these babies are." Joe guided me up and down the length of the station, touching each pump with a personal stroke, telling me over and over again that each could pump 75,000 gallons of water a second.

The pumps were not operating when I was there, save for a relatively tiny vertical pump capable of handling the trickle of water coming into the reservoir. It must be something to hear and see these monsters at work. Joe kept telling me to keep alert for a big storm; that's when I should come back. He wanted to show me the pumps at work then.

Serious excavation was taking place at one end of the pumping station. I asked Joe why. He explained that two more pumps were being installed to handle the ever-increasing runoff. He sounded disgusted. "There's more rainfall, more shopping plazas going in, and the city keeps sinking." He spat it all out with astounding succinctness. There *is* more rain. Between 1900 and 1978—over three-quarters of a century—storms that delivered more than five inches per hour pummeled the city twenty-two times. But after the 1978 storm—the one that introduced me to the ferocity and rapidity of flooding in New Orleans—the pace quickened; in the past two decades the city has endured ten storms that have each dumped equal or greater volume. Over the same two decades, more and more of the city continues to be paved over, as malls clone each other and as their parking lots blend into a vast, impervious asphalt blanket that forces rainfall to rush down storm drains and inundate the system rather than be absorbed by the ground.

Toward the end of a three-hour tour of New Orleans's drainage system, Joe and I ended up at the largest pumping facility in the world—DPS No. 6—Drainage Pumping Station No. 6 at 345 Orpheum Street. This giant receives storm runoff from most of Orleans Parish and some of Jefferson Parish. Its fifteen pumps can move over *six billion* gallons of runoff per day to Lake Pontchartrain. Wood designed most of them, two with fourteen-foot-diameter iron pipes housing the impeller and each powered by a twenty-foot-diameter motor coil that you don't want to get too close to when the pumps are running. The aura of electricity surrounding them can give you a "severe tingle," Joe said. They can also grind you to a pulp.

We went to the lakeside edge of the pumping station to look down the two miles of open canal toward the lake, where the six-billion-plus gallons of water are sent after passing through the pumps. Anna St. Romaire, the station's operator, who has worked for the S & WB for eighteen years, accompanied us, mostly because there was not much to do in the station. The complex of dials and LEDs in the control room rotated and blinked in synchrony with the bright day. The vast concrete floors were swept and the huge machinery hummed in readiness for the next deluge, though no appreciable rain had fallen for the past week, and none was forecast. Anna said that she had guided a Chinese delegation through the station yesterday, followed by a horde of school kids. Today she needed a rest.

Oddly, the canal to Lake Pontchartrain was brimming with water. It had risen to just two feet beneath the struts of a nearby bridge. There was a north wind blowing, though gently at fifteen miles per hour. Joe told me that the canal was high because wind had pushed water into it from the lake. In 1998, the leading edge of Hurricane Georges pushed so much water into the canal that it had lapped the bridge struts and came within a few inches of overflowing the canal's concrete rim. Fortunately, Georges, like Betsy, let loose only about four inches of rain, enough to get the pumps going but not to create an emergency.

But enough to create the potential for a serious dilemma. When wind pushes water into the canal from the north and east, and pumps are working to push water in the opposite direction, out of the city and into the lake, the pumps end up doing extra work, reducing their overall effectiveness by 20 percent, Joe estimated, and preventing easy flow of floodwaters to Lake Pontchartrain.

Anna poo-pooed my concern over the implications of this bottleneck. In her eighteen years on the job, she assured me, the sort of hurricane that pushes winds over the lake never produces much rain. So not to worry. I asked Joe what would happen if, just suppose, winds from a powerful hurricane which carried a lot of moisture did come in over the lake and pushed water over the levees into the city *and* at the same time dumped thirteen or so inches of rain on the city, as did Frances, a mere tropical storm, in September 1998. What then? Joe turned to me, looking kind of startled by the question. He hesitated and then said, his voice firm and clear: "We (S & WB) only deal in storm runoff. We're not designed to handle hurricane floodwaters, too."

Frank Mineo concurred in a different way. As the planning

engineer for the Orleans Parish Levee Board, he is supervising the installation of floodgates on the five storm canals that drain water from pumping stations into the lake, a decision prompted by Betsy's flooding in 1965, but only acted upon in 1999. The project is actually an Army Corps of Engineers undertaking where feasibility studies that dwell on cost-benefit ratios, authorizations, and implementation can occupy a good number of years. The idea is that when the construction is finished—but that will not be until 2005—the gates can be closed if water rampages up the canals, threatening the pumping stations. It sounds like a good idea, but there's a problem, which Mineo broached to me in the same embarrassed way that Joe Puglia had. The project design—as the Corps terms whatever they are building—calls for the floodgates to withstand the predicted eleven-foot maximum storm surge of a fast-moving Category Three hurricane, with maximum winds of 130 miles per hour. Well, Betsy barreled in, carrying 150 miles-per-hour winds and Camille in 1969 was just plain terrifying at over 200 miles per hour.

That people in Louisiana are not preparing themselves for The Big One worries Frank Hijuelos, who heads the Office of Emergency Preparedness for the city of New Orleans. The partial evacuation of the city, as Hurricane Georges swirled out of the Gulf in 1998, was not smooth, to put it mildly. The Superdome, converted from a sports arena to a giant shelter, took on the atmosphere of a holding pen. Little was provided for the thousands of people who sought refuge in its cavernous interior. They had to wait in the gloom, cut off from even being adequately updated on the storm's progress. Fights broke out over vending machines, seats were slashed, the toilets clogged. Those who could evacuate by car considered themselves luckier. That is until they found themselves on Interstate 10, the major route out of the city. Subsequent stories of families traveling one mile in six hours have become part of the city's evacuation lore. And the efforts, it turned out, were for nothing, as the hurricane veered east and hit the Mississippi coast.

Tens of thousands of people stayed put, either gambling that they could ride out the storm or because they had no way to get out. In the end, they were the fortunate ones, at least this time around. But many were elderly or disabled and had no choice. Hijuelos estimates that over 100,000 people in New Orleans's poorer neighborhoods have no access to cars. It is these people, in addition to those inured to hurricanes that never seem to strike New Orleans, that really worry Hijuelos. "This is a very conservative city," he told me in his city hall office. "People here don't like to move. A frightening number of people just don't get it." The "it" refers to the fact that New Orleans just has been plain lucky. What many people don't realize is that the luck is bound to run out. And when it does, tens of thousands of people may find themselves trapped in a sunken city.

Even worse, the city may not exist as we know it. The Army Corps of Engineers is well aware of that possibility. In fact, the Corps, the National Weather Service, and the Federal Emergency Management Agency (FEMA) have pinpointed three areas in the country as most prone to disaster. One is the West Coast, under threat of massive earthquake damage; another is the Madrid earthquake zone in Missouri; the third is New Orleans and its vicinity.

As a result of a Category Five (winds above 155 mph) or a slow-moving Category Four hurricane (winds between 131 and 155 mph), life in South Louisiana as it has evolved over the past 300 years could abruptly end. Jay Combe, chief of coastal engineering for the Corps's New Orleans District, has a map showing the extent of the possible flooding. In the worst case, the entire metropolitan New Orleans area, including the New Orleans International Airport, will be under twenty-five feet of water. Combe, speaking in measured tones about the predicament facing New Orleans, says that three days before that event occurs, he will be sure to take himself and his wife as far north as Vicksburg, Mississippi, and he advises everyone else to do likewise.

Three days after the hurricane passes, the waters will have receded to the top of the levees that surround the city. Their lowest height is fourteen feet, along the New Orleans-Metairie border, though rail crossings leave some open gaps in the city's protective wall. Then the "un-watering" of New Orleans can begin, as

Combe calls the process, which has been elaborately conceived by the Corps.

Punching holes in the levees is the first step in "un-watering." How such an undertaking will be accomplished is solemnly intriguing. The Corps's New Orleans District, responsible for the massive levee-building around the city, will not be able to assume the task of breaching them. District headquarters will have been abandoned. In fact, all municipal functions will have ceased even before the hurricane's onslaught—pumping, transportation, electricity, communications of any sort. Whoever punches holes in the levees will not have a detailed knowledge of their locations, nor of the pumps. In preparation for this eventuality, Combe, "the guy who makes up nightmares," as he calls himself, has been working on what he labels a "notebook" for over a year. It is actually a four-inch-thick loose-leaf binder depicting the location of everything needed to begin to resuscitate the city—levees and their materials, pumps, rail lines, drainage canals, generators, etc. In a way, his notebook is a legacy to those who come upon the wreck of New Orleans.

In all likelihood that will be the Corps's Memphis District Office. Personnel and equipment from that office will break through the levees using the information in the "notebook" (which will have been transferred to computer) as a guide. Water trapped in the city for the past three days will gush out of these crevasses, carrying with it houses, cars, and corpses. The 60 percent of the city below sea level, however, will remain flooded in five to eight feet of water.

The next step is to get the pumps going, which means, of course, restoring electricity to them. Combe predicts that it will take at least a week to begin pumping the water out, a year before New Orleans is able to again function as a place to inhabit, many years, if ever, before it becomes a viable city again. "When I told my bosses that it would take a year to get the city going again," says Combe, "they said that that was too long. They said I should make it six months." He gives a shrug; whatever the stated time period, it's only another guesstimate.

So massive is the threat to New Orleans that little is being done to guard against it. The only real accomplishment since Georges's threat is embarrassingly obvious—a plan to ease evacuation by opening all lanes of Interstate 10 to outgoing traffic. So while Frank Hijuelos worries about New Orleans's paralysis in the face of a hurricane, Joseph N. Suhayda, a Louisiana State University engineer who directs the university's Louisiana Water Resources Research Institute, is offended by it. Suhayda looks like everyone's favorite uncle—tall, bespectacled, slightly balding and with a totally friendly face that masks his dire message, one with which a growing number of people agree: that virtually everything the state has done to protect itself from future ravages of nature is not enough. Just seeing the upset look on the face of this family-looking man invokes an obligation to pay attention; hearing his message encourages a whole new way of thinking about South Louisiana's future.

I first met Suhayda in a large, mostly empty conference room in the Doubletree Hotel in New Orleans. We were both waiting for a conference to kick off. It was August 17, 1999, a date of no importance except that it marked the thirtieth anniversary of Hurricane Camille's hellish battering of the Mississippi coast with winds estimated at 230 miles per hour and thirty-five-foot storm surges that flattened the town of Pass Christian just west of Biloxi and fifty miles east of New Orleans. Camille is the second Category Five hurricane to have struck the United States; the first was in Florida in 1935. Again, New Orleans had been spared.

The conference was held to commemorate Camille and to warn of future hurricanes to come. We introduced ourselves, chatted first about Indiana, where he's from, and then about New York, where I live and, of course, about how much more vulnerable New Orleans is than both of those places. I didn't realize exactly how vulnerable until the following morning.

Suhayda had worked up a computer simulation that scares a lot of people. He ran it on a big screen in the conference room and you couldn't hear anyone even draw a breath as they watched what happens as a computer-simulated Category Four hurricane sweeps in over the city from the southwest, moving at ten miles per hour. Little white arrows representing a 130-miles-per-hour wind fly across the screen, curving northeast. Their counterclockwise rotation marks the wind's intensity. It pushes water into Lake Pontchartrain from the Gulf, and you think you are watching the beginning of a bad Hollywood disaster film, the part where a scientist has created a computer simulation of what might happen. As the storm moves northward, rotating on its axis like one of Wood's pumps, the volume of the lake—a vibrant blue on the screen— swells like a jellyfish bloating under the sun. At the same time, a twenty-seven-foot-high storm surge moves across the decimated marshes southwest of the city, tops levees, and crashes through the malls, strip developments, and expanses of subdivisions on the west bank of the river south of and opposite the city. The land, green on the screen, transforms to blue as floodwaters bury it.

The eye of Suhayda's simulated storm takes a day and a half to pass just to the east of the city. When that happens, its winds are out of the north, blowing directly across the lake and pushing its entire bulk of water against the levees that protect the city. The levees crumble under the force against them, as if a giant bulldozer had pushed them down. Water gushes everywhere, drowning the city. The flooding is most severe in low-lying sections, the flatlands between the river's and the lake's natural levees. This is where most of the pumps are located. And now, of course, the pumps are entirely underwater and inoperable. What now? How do you pump water out of a giant saucer when the pumps don't work?

Suhayda doesn't know, but over the past year, he has made it his business to inform Louisianans that the efforts being made to protect their state are fantastical in their inadequacy. As evidenced by the breathless silence during his presentation, a silence of scientists, policy experts, politicians, and media, his volubly articulated fears are in concert with others' silent nightmares. Beneath the hype about how well the state is squaring off against the future, few people are willing to come to terms with the enormity of the problem. New Orleans may have in place a venerable drainage system that thrills school children and visiting engineers alike—a barrier of levees that has tamed the Mississippi, a battalion of worried citizens and scientists who move sand from one place to another and plant marsh grass in dying wetlands—but that's not good enough for Suhayda.

"We are hung up in a mode of operation that is not having an impact," Suhayda tells me a few months later in his computer-studded laboratory at LSU. "It's a Vietnam syndrome—maybe if we do just a little more, we'll win."

Though Suhayda has displayed his hurricane-devastation simulation many times around the state—to policy planners, levee boards, the Army Corps of Engineers, the state legislature—and though people continue to go all quiet when they see the flooding—the novelty is used up. What can you do when you see the city filling with water like a bathtub except watch in horror? Yet this possible horror has not translated into action, save the creation of Jay Combe's notebook.

Suhayda, though discouraged, is not downed. In a healthy reaction to his frustration, he is developing a new computer simulation. "My mission now is to bring to the fore an objective—what do we want to do to save ourselves and our economy?" he explains rather cheerfully. This is a different kind of question than has been asked before. It articulates an objective and then strives to reach a goal, quite different from merely attempting to protect what is left.

The new simulation employs a representation of the state's wetlands as they appeared in 1830, when they covered far more expanses than the present Swiss cheese look. Then, a trapper could walk from the Mississippi east to Houma and not fall into Barataria Bay, because back then, the bay was a tiny indentation in the coast. Overlaying this perspective are progressive sequences of deterioration of the wetlands up to the present. Barataria Bay looms larger and larger; barrier islands diminish and break apart; interior ponds suddenly appear as if asteroids had assaulted the landscape. In its way, the sequence from stability to ephemerality is as terrifying as the flooding of New Orleans, perhaps more so with the realization that the flooding would be much less extensive if the marsh were intact. No marsh means big storm surge.

As Suhayda gives me a private showing of his work in progress—his "disaster already made," as he calls the decimation of the coast—he tells me about a disconcerting discovery he has made during his research. It is that all the work of the Breaux Act's task force in its efforts to save barrier islands, to plant marshes, to restore hydrology, to create wetlands from dredged material, all this work, which has cost the federal government and the state some $300 million over the past ten years, has accomplished absolutely nothing except to slow down the loss of wetlands by 15 percent per year. In other words, though the task force's mission is to restore wetlands and barrier islands, ultimately nothing is being created, and the continuing loss is estimated to total another one-thousand square miles of wetlands by 2050.

What has to happen, Suhayda believes, is to find a solution to the problem that fits its scale. But before that can happen, people, whether they be fishermen, government officials, politicians, or environmentalists, have to determine what they want. That lack of determination has been the undoing of present efforts. "We want to save the coast?" he asks, his voice full of rhetoric as his technology hums and blinks as a sort of passive encouragement. "OK. But by how much? Back to the 1830 coastline, 1900, 1940? Unless you have a threshold, all your efforts will do no good."

Why Louisianans have never made up their minds about the extent of their willingness to save their environment puzzles Suhayda. He claims it would be very easy, actually, to restore the wetlands surrounding Barataria Bay and squeeze the bay down to the very reduced size it was 150 years ago, adding tens of thousands of acres of marsh to the coast. He wonders why huge projects of the distant and near past—like building cathedrals in fifteenth-century France, as one of his examples—were accomplished, and something of such relative simplicity as saving a marsh can't be accomplished. "Where is our center?" he muses, then shrugs and concludes his thought: "I don't see an obligation here to restore the coast. Well, society does what it does and has to deal with the consequences."

Suhayda is not alone in his frustration. Virtually everyone who worries about the demise of the state's wetlands and the long-range implications shares his frustration. The disregard of the sculpting of this land has resulted in the unintentional emergence of a monster. Suhayda's simulations articulate, as nothing else has, both the reach of the monster's rage and helplessness to rein in its destructive powers. The vivid onslaughts of wind destruction and marsh destruction on his computer have forced a realization that time is running out. Whether from a hurricane or from ecosystem collapse, a crisis is coming.

From this awareness of approaching crisis, progress to stop it is finally making some headway. The crisis that looms in Louisiana is a novelty for both the state and the country. No one anywhere has experience in stemming such a disruption to a natural system. This is not a case of cleaning a polluted river, saving a forest, stopping a road. The Louisiana wetlands offer the challenge of pacifying a monster artificially created. Particularly difficult is the necessity of discerning the natural from what *appears* to be the natural. So integrated have human encroachments become on the landscape—oil field canals, levees, jetties, spoil banks—that they are often considered part of the ecosystem. And even more complicating, their dismemberment would create an entirely new, and unfamiliar, natural system.

It is the custom of people in times of uncertainty to rely for guidance on the known. In Louisiana, the known remedies—the grass planting, the dune building, the little freshwater diversions—have been tried. The traditional bureaucracy that administers them has proved unwieldy and ineffective. The progression toward this conclusion is probably the most optimistic knowledge that has come out of Louisiana's struggles. It puts Louisiana against the wall. And forces a new way of thinking about its survival.

That new way is just beginning, an experiment for all the country to learn from and to witness.

Chapter Nine

Jack Caldwell's Big Day

Rectangular tables are arranged in an open square in a conference room of the Army Corps of Engineers' district headquarters in New Orleans. The room is square. Hues of gray predominate—carpet, chairs, curtains, walls. The curtains are drawn. If they were pulled, the view of the Mississippi River on a glorious December day would be grand. But this is not a time for atmosphere. This is the meeting of the Technical Committee of the Coastal Wetlands Planning, Protection and Restoration Act (CWPPRA) called to advise in the final selections of wetlands restoration projects proposed for the ninth year of funding under CWPPRA, known also as the Breaux Act. There are thirty projects under consideration today, whittled down during the year from sixty.

CWPPRA task force reps stake their turf; voluminous papers extracted from folders, charts unfurled, and spreadsheets scattered—all territorial table-markers. Corps officials take the table facing the public. The Department of Natural Resources (DNR) is just to the right. With considerable foresight, its three representatives spread themselves out so they occupy a corner where two tables meet. More clout that way. And just to their right is the governor's office itself, at least in the form of Len Bahr, Governor

Mike Foster's executive assistant for coastal activities. Natural Resources Conservation Service comes next, then the EPA. On the left side are National Marine Fisheries Service (NMFS) and U.S. Fish and Wildlife.

The audience—maybe twenty-five individuals willing to sit through an imminent several hours of minor political drama—mill about still unoccupied chairs near the front of this somber room. They look about with airs of expectation. There's considerable back-and-forth between some of these spectators and the agency reps, last-minute little deals, pleas, suggestions, all wrapped in a mantle of grace and jocularity that is so particular to South Louisiana, even here deep in concrete bureaucracy. When all is said and done, there's still family to go back to, fish to catch, and stories to swap, in that order.

I decide to be one of the spectators in hopes of discerning through this event, and other similar ones, some of the minutiae of Louisiana's effort to save itself. People who know what I am doing tell me that I will surely succumb to boredom within minutes; others advise me that nothing can save the state from disappearing. A few think the entire CWPPRA process and projects no more than absurd child's play—so insufficient before the magnitude of the Gulf's encroaching waters as to give the proceedings a sort of Mardi Gras bombast.

Before the meeting begins, I run into Jack Caldwell, secretary of DNR, and the highest-ranking person in the room. He is still wearing the same jovial southern-hospitality smile he sported when he was consuming fried softshell crabs at the Kajun Sportsman in Port Fourchon and sparring with Sue Hawes. He tells me he is "so glad" that I have come. He advises me that something "very interesting" is going to happen, "something that is going to change the way we do these projects." He asks me out to lunch after the meeting.

The lead Corps rep calls the meeting to order and announces that the thirty candidate projects have suddenly been pared down to twenty (as a result, one supposes, of the invisible negotiations that lend the process an aura of mystery). Initial engineering and design costs for these twenty projects, the rep says, will come in at

$20 million. The figure has been calculated on a strict cost-benefit analysis, one dollar of cost equaling one dollar of benefit. The hoped-for benefit is beach stability, marsh preservation, flood control, etc., derived through the uncertain but accepted process of running guesstimated figures through a complex formula with the weighty name of wetlands-value analysis.

Most people familiar with Army Corps of Engineers' thinking understand the shortsightedness of cost-benefit—the kind of mentality that has produced arrow-straight channels like the Mississippi River Gulf Outlet—and an affection for "hard projects" like rock jetties and concrete bulkheads, rather than sand-dune creation and marsh grass planting. Cost-benefit analysis does not pay much attention to long-term outcomes; the Corps's mandate, after all, is flood prevention and navigation maintenance, and it has done an admirable job in these two areas. And cost-benefit is easy to understand, easy to explain on a spreadsheet, easy to sell to politicians and public, as straight and clear as the MR. GO. itself. So, it has remained the modus operandi of the Corps and of these meetings.

The notion of cost effectiveness, another buzzword for cost-benefit, is so comforting that many CWPPRA agencies rely upon its concept, oblivious to the thought that the so-called benefit, or effectiveness, is often concocted with no idea as to what damage the next hurricane or winter storm might bring. Plus the fact that a benefit in the form of flood control may indeed prevent flooding as an immediate benefit, but over time it may kill a wetland.

So, it brings some relief when the Corps rep announces that the most favored project is a diversion of Mississippi River water—tiny, at only four thousand cubic feet per second—through the Bonnet Carré spillway[1] as a way of bringing silt into Lake Pontchartrain to help

---

[1] Historically, the natural levee at Bonnet Carré, thirty miles upstream from New Orleans on the east bank, has been weak. In 1871 and in 1874, the river broke through the levee, pouring water into Lake Pontchartrain. In 1937, the Corps built a gated concrete sill in place of the crevasse. This spillway, designed to allow up to 250,000 cubic feet per second to flow from the Mississippi in flood to relieve pressure on New Orleans, has been opened numerous times, most recently in 1993.

build up the deteriorating marsh. Almost four hundred acres of marsh have disappeared from the edge of this lakeshore since 1930 due to canal dredging for the old Bonnet Carré Oil Field, and other human messing about. If something is not done soon, the Corps predicts that the lake will start lapping at the embankment that supports the main rail trunk that carries freight and passengers from New Orleans to points north—important points like St. Louis, Chicago, and Detroit. The plan, proposed by the National Marine Fisheries's CWPPRA task-force reps is simple, as simple as nature. Unfortunately, it is also so small, so unobtrusive in the face of the onslaught of loss, as to be a mere blink of the eye.

Other candidates in slightly less favor include: building control structures like gates across canals to keep salt water out of freshwater marshes; dredging channels to allow fresh water to flow into brackish marshes; fabricating rock breakwaters to break storm surges likely to wash away barrier islands. The acreage guesstimated to be saved or created by these projects is amazing: 25,000 acres from a freshwater diversion into Lake Maurepas, 8,400 acres from a sediment trap south of Venice, and stabilization of six hundred acres of salt marsh on the Chandeleur Islands. The irony in these dreams is the horrible statistic floating about local management circles here—that no matter how successful these projects are, the state's hemorrhaging of wetlands is being reduced by only 15 percent per year. In other words, the annual loss of twenty-five to thirty-five square miles of wetlands is dwindling by around only three to five square miles.

The vastness of loss in Louisiana places the entire wetlands issue on a scale unimaginable in other parts of the country, where an acre of loss, an acre of gain, assume sacramental status. Shea Penland once told me the manner in which people over in Galveston, Texas, idolize marshes that have been formed from spoil material dredged out of Galveston Bay. People go out on the bridge that crosses the bay and rhapsodize over their little artificial hummocks below. They admire the amount of "interspersion" their tiny marshes possess, which means the indentations at their edges. The more involuted the perimeter, the greater the nurturing value for minnows, crabs, and shrimp, all the way up to great blue herons. Louisiana is so huge in marsh areas that that kind of thinking just seems out of line, like contemplating the value of a mud puddle.

Jack Caldwell suddenly harrumphs and churns in his seat like a tug's wake. The room turns quiet. Attention garnered, he announces that the agency he leads, the DNR, is introducing what he calls "a big change in philosophy" in the way restoration projects are chosen under the Breaux Act. No one knows what he is talking about but it is as if the room and all the people in it have become one living entity, inhaling and emitting breath as one, listening as one, watching something entirely unusual unfolding, something historic. What the people in the room are witnessing is a moment in environmental history. It is a moment of acknowledgment, of desperation, of panic, of resilience, and of hope. Caldwell's "big change" is not only a reluctant admission that the CWPPRA projects, at a price of $300 million over the years, have accomplished very little against the natural forces that are eating at the coast like an autoimmune disease. It is also a recognition, perhaps a final recognition, that if the state with its present 3 million-plus acres of wetlands is going to be saved, the rescue must be designed to fit nature rather than political inclinations.

There had been considerable talk in late 1999 about the way money was being spent for restoration projects. A five-part series in *The Advocate*, the Baton Rouge newspaper, pointed out that many of the projects were ineffective. Mike Dunne, the paper's environmental reporter, also drew attention to a mammoth fiscal inefficiency: much of the money allotted to the projects was sitting in a bank, rather than being put to use for other coastal salvage efforts. Since the beginning of the Breaux Act, each project selected received a budget that included design, engineering, and maintenance costs over a twenty-year period—the supposed lifetime of the project. So, while many projects were in the design stage, millions of dollars for their eventual construction and maintenance were idle. Now, at the meeting, Caldwell castigates the lack of cost-effectiveness in one breath and, in the next, launches into a speech about cash-flow management. And in the following breath he says that the DNR, in concert with apparent ally Len Bahr from Governor Foster's office, proposes to pare down the Corps's list of projects from twenty to seventeen and reduces the present budget from $20 million to $16 million.

The way things are going to operate from now on, Caldwell says (though he is entirely lacking in fiat), is that money will be allotted to each project for design and engineering only. After these initial stages, progress and value will be reassessed and additional funds ponied up, but only if warranted. The object is to free up the excess funding that has been restricted to each project from the outset. And the purpose behind the change is increased flexibility: if a project looks like it is not going to work in its design stages, the design can be changed, or the project can be dropped altogether and the money diverted elsewhere.

In many ways, the quest for increased flexibility tries to allay the sense of increasing frustration at the ongoing wetlands loss. Over the years, too many projects have been approved and funded, then bogged down due to poor management or poor design or opposition from oystermen or landowners. "The reality is that there is slippage in the number of projects getting going," declares Randy Hanchey, Caldwell's right-hand man, who seems to have assumed the charge. "The point is that the marsh is dissolving."

Logical as Caldwell's proposal appears, logic does not immediately hold sway. A way of doing business is being challenged and that calls for rebuttal. The U.S. Fish and Wildlife Service rep leads off with an argument for stashing all the money away so that if a really good project comes along—the magic bullet to save the marsh—there will be sufficient funds to carry it out.

Len Bahr goes berserk with exasperation. "The magic bullet never seems to appear," he dashes out with terrible truthfulness. "We can't wait. We have to act now."

Rick Hartman, the rep from the NMFS, protests that the change to cash flow will interfere with the elaborately conceived formula to predict cost-effectiveness that the CWPPRA agencies worked out long ago. A couple of other reps at the table grumble about questionable cost-effectiveness, uncertain environmental benefits, this and that—more evidence that the CWPPRA projects are not working.

Then a strange thing happens. Hartman does an about-face and makes a motion that the seventeen projects and various "demonstration projects"—meaning that they are of uncertain benefit and perhaps a waste of money—be approved. The motion makes his protests of two minutes ago appear mechanical, as though he had been driven by an ethic deeply ingrained in his organization's bureaucracy. It must have taken some courage to make that sharp turn. The projects are quickly voted on and accepted.

Caldwell glows; but this is not the end of the DNR's work. This is just the beginning. The DNR has created an elaborate menu of offerings in recognition of CWPPRA's failure. Caldwell sits back and Bill Good, another of his lieutenants, who, like Randy Hanchey, imparts a professorial/businessman mien, holds forth on—rather, barbecues—the selection procedure for the past nine years, based as it has been on the predilections of each agency. He calls it a "Model T," and gives the new proposal, about to be articulated, the unfortunate appellation "Model A," rather than, say, "Thunderbird," or "Explorer." "Rather than working from a piecemeal fashion up, let's work from the big picture down," he lectures.

Again, stillness steals over the room, absorbing the inevitable throat clearings and paper shuffling. Change looms large in the air, a change that, as change does, threatens the status quo and makes people edgy. "Let's address the greatest need first; let's prioritize projects based on ecosystem need," Good suggests. The entire coast, he announces, should be segmented into hydrologic basins, vaguely conforming to the various deltas that the Mississippi has deposited over the past seven thousand years. One includes Lake Pontchartrain and its wetlands; another is Breton Sound, the present Mississippi's delta, and Barataria Bay, and so forth. And each of these regions should then be divided up into "ecosystems," as Good erroneously calls them—really, specific drainage areas and marshes in need of help.

Randy Hanchey fills the hovering silence: "The connection between specific projects and the big picture is missing. This plan will get us away from the agency emphasis. We need to think in terms of watersheds."

Sue Hawes, largely quiet up to this point but still casting a sage-like aura for the Corps, solemnly utters: "I think everyone agrees that the nomination process (for CWPPRA) is broken."

But not everyone agrees. The EPA rep worries how each agency's "interests will be represented" under this proposal. To which Randy Hanchey counters: "Let's back up and look at the watershed." Rick Hartman worries again about cost effectiveness, to which Len Bahr responds: "Look, everything is getting cut up right now with our present project nomination system. We have no regard for ecology."

But it's too much for the agency reps to handle, too much to ask them to give up their pet ideologies and projects here and now. Murmurs go around the table to the effect that such a radical departure requires deliberation. Someone makes a motion to table the DNR proposal, a motion that is voted for so fast it can scarcely be witnessed. The meeting quickly dissembles into reports that garner a fraction of the previous collective attention. In the middle of one of them, Caldwell packs up his papers and bustles over to me like a minor Rumpole and urges me to share with him "some of the best oysters in town." "They're charbroiled, ya see, and no one else in Nawllins does 'em like this."

A half hour later, somewhere deep in the malls of Metairie, we are seated at Drago's, scoffing down oysters on the shell grilled in garlic and butter to create a blend of taste sensations that is entirely new to me. The place reminds me of a Mafia restaurant-office and at three in the afternoon, we are its only customers. Caldwell is resigned, not happy about the meeting's outcome, but reluctantly appreciative of the trauma the day's proposals have engendered. "These agency people are real scared," he tells me. "They need numbers to hide behind like cost-effectiveness and their wetlands value-analysis. We're trying to get away from this stuff but I am not a windmill tilter. I know that change is going to be gradual."

While Caldwell, moving on to a substantial bowl of gumbo,

has mixed feelings about the meeting's outcome, he is at the moment decidedly pessimistic about the bigger picture—efforts in motion to extract sufficient funds from the federal government to enable South Louisiana to save itself. The quest for national attention that the struggle for Louisiana wetlands preservation deserves, but does not always receive, is exhausting.

Two weeks ago, Caldwell had been much happier. Down at Port Fourchon for a dedication ceremony of some minor CWP-PRA projects, he was crowing over an article that would shortly appear in the *Boston Globe* that outlined Louisiana's plight. "Can you just imagine," Caldwell glowed to a circle of minor pols and agency types, "The *Boston Globe*! This puts our situation right on the national map." It was an enthusiasm that said as much about Caldwell's naiveté as it did about his optimism.

But today, Caldwell is down, feeling diminished by the agency people's insistence on cost-benefit ratios and other examples of paralyzed thinking. "Tomorrow, who knows?" Caldwell keeps saying how he hopes that *The New York Times* will do a story on the state's coast. "In fact," he tells me, as he gets into a battered sedan with a FOR SALE sign in the rear window, "they sent a reporter down here not long ago. Who knows?"

A couple of months later, I am eating lunch with Len Bahr and Randy Hanchey in Mat and Naddies Restaurant, where the best crawfish cakes in the whole world are served. It's after another meeting, this one dealing with issues over the proposed closing of the Mississippi River Gulf Outlet. I ask them about that December meeting, the obvious tension, the reluctance of the federal agencies to see the merits of cash-flow management of the CWPPRA projects and to understand the strengths of an ecosystem approach. They both nod and Bahr comes out with a refrain that I am hearing with frightening frequency: "We are on the *Titanic* here. It looks okay now, it's listing just a little so far and all we are doing is rearranging the deck chairs."

Their frustration at these endless rounds of meetings is high.

They complain that the different CWPPRA agencies have never worked together—that they each have their own cultures and interests. The U.S. Fish and Wildlife contingent, for example, is only interested in marshes, not barrier islands. The Corps does not like barrier islands either, says Hanchey, who is a former Corps employee himself. Barrier islands mainly offer what he calls "symbolic value," meaning that they create a gradual shift from ocean to land, that they mildly buffer us from the ocean, that their beaches provide memories of childhood summers, that their lighthouses warm us even in winter.

But given the competition and cultural provincialism of these conservative bureaucracies, what would happen if Louisiana suddenly found itself with a substantial check each year for coastal restoration? A bill introduced in Washington recently, the Conservation and Reinvestment Act of 1998, had opened the possibility that Louisiana might stand to receive some $300 million a year earmarked for the coast. Both men sit upright at the table; it is a question that concerns them and a lot of other thinking people in the state. After a second's reflection, Bahr offers: "CWPPRA would be thrown out. We would have to design a whole new way of doing things."

A new way at looking at things that a shot of money would have to precipitate might be the best thing to happen to the state for a long time. If it came from the federal government, from the country's taxpayers, it might exert a certain amount of efficiency and innovation in ways of dealing with the coast's disappearance. The place of nature, in this case Louisiana's wetlands, far beyond Louisiana's borders, would have to be considered. It would force the rest of the country to take a look at Louisiana's crisis. In many ways, the Louisiana coast has to do with more than Louisiana; it is a national treasure that contributes to a way of life and thinking that is as visible—if you look—in New York as in Albuquerque, New Mexico. While Louisianans may be accused of petty bureaucracy and cost-benefit thinking, the country as a whole has yet to understand that these wetlands are as important to everyone as the Grand Canyon or redwood forests.

Chapter Ten

Lessons Learned

At the annual meeting in February 2000 of the Louisiana Landowners Association—a group of mega-acreage marsh and swamp owners—the frustration and tension are palpable in the Embassy Suites conference room in Baton Rouge. You meander through a baffle of little waterfalls and winding streams in the atrium lobby. You watch the kids ogling the wavy-tailed goldfish. The luxuriant vegetation in the atrium is dreamy. Then you walk into the conference room and wham! the air is thick with grumbling landowners. They look well-mannered, but you wonder if they have machetes under their jackets. They're not sitting, listening to the speakers up on the stage; they're standing and swaying and fidgeting, like they're about to leap up and start swinging out of frustration. Alex Plaisance, the association's leader and as big a landowner as any of them, is telling them something about New Orleans being the next Venice. "If we don't do something now, we will lose everything," he pleads. Someone in the audience yells out: "We should all be buying up Arkansas so we can move there when Louisiana is gone."

They get really restless when Jay Gamble, the Army Corps's educational outreach coordinator, talks about an eighteen-monthlong feasibility study to determine how to spend $3.5 billion in efforts to save the coast.[1] These people are sick of studies. They are tired of waiting, and their almost synchronized chatter amounts to "not *another* study." The exasperation does not diminish when Len Bahr, up on the stage and speaking on behalf of Governor Foster, announces that "things are just not moving fast enough" and complains about the "tremendous loss of parish revenue as the land goes away."

By the time a fellow named Mark Davis gets to speak, the landowners are spent, sluggish in the old belief that nothing but talk is being expended to save their land. But there's something chemical wafting through the room. It has to do with Davis, a tall, gangly guy, nerdy and sharp, appearing so obviously not from these parts with his rapid-fire gesticulations, flat northern accent, and darting eyes. He's from Chicago, a lawyer who used to work in tax and transactions. He came to Louisiana by accident. His former wife, also a lawyer, received a clerkship with a Louisiana judge. Davis came as a trailing spouse and found a job as counsel to the Lake Pontchartrain Basin Foundation. Now, he directs the Coalition to Restore Coastal Louisiana—CRCL—a grassroots organization that has won over all sorts of people for its even-handed concern for the future of the wetlands.

Davis picks up with the bad news but gives it an inspirational twist. He says beating coastal loss is a "survival challenge." "The real victim here," he goes on, "is loss of economic opportunity." The landowners are more wakeful now. There's something infectious about Davis's no-nonsense, no-politics approach. Maybe it's

---

[1] Early in 2000, the state and the federal government, via the Army Corps of Engineers, each agreed to contribute $3 million toward feasibility studies that hopefully would lead to ways of dealing with coastal erosion. The studies, requiring three to four years, would focus on the benefits of marsh creation, barrier island restoration, freshwater diversions, and hurricane protection. All these areas have already received voluminous study. What was new in the state-federal agreement was the equal expenditure. That the state agreed to foot half the bill is hoped by state officials to signal to Washington a commitment to get serious about the coast's demise.

because he so obviously is not from these parts, "We have to make sure that what we do with our wetlands is Louisiana's proudest story. What hurts so much is to hear someone who has flown over the marsh ask: 'How could *you* have done *that*?'"

What can landowners do? Simple, says Davis, with a slightly scolding tone. Don't sit back and complain. You are powerful people. Make sure that wetlands loss gets high-profile attention. Get involved. It's a humbling message; it's no secret in Louisiana that a lot of big landowners watch their land vanish, file their complaints, and shrug their shoulders. Maybe the problem is that they have *so* much land. Another problem is that a large percentage of the coast is owned by individuals and corporations in New York, Chicago, and Atlanta.

Getting involved is a new message in South Louisiana, part of an emerging tactic to present a populist front against the continued desecration of the coast. It implies that the state and federal bureaucracies do not, perhaps cannot, do the job. It a cry from and to South Louisianans that the coast will be saved only if everyone, from oilmen to oystermen, understands that the state's future is based on the continued existence of the wetlands. It's a recognition that preservation of an environment—and here's the lesson for other parts of the country—is only a scrap of the picture; the complete view is in the acknowledgment that environment and people go together. Both have imprinted this land. To appreciate uninterrupted nature is fine; the fact remains that we human beings have as much right to benefit from that nature as nutria have a right to eat grass. The difference, only recently recognized, is that humans can temper their influence. Nutria cannot.

It's taken a long time to get to this point—that wetlands survival in Louisiana is everyone's issue—over a quarter of a century. With the completion of eighteen reports written between 1970 and 1973, the scope of the problem was laid bare and remedies offered. Entitled *Environmental Atlas and Multi-Use Management Plan for South-Central Louisiana*, the final reports served five functions: they identified the extent and location of land loss, pinpointed the causes, assessed the characteristics of the area's unique ecosystem, recommended the installation of freshwater diversions, and advised the development of a multiple-use planning approach. Everything seemed to be in place for action. But there was no action.

Since then, the Louisiana legislature has established dozens of agencies, task forces, and focus groups, etc., which have studied aspects of coastal deterioration. The bounty of organizational names, many still in existence, is numbing. Here is an abbreviated listing: the Louisiana Advisory Commission on Coastal and Marine Resources; the Louisiana Coastal Zone Management Programs; the Coastal Protection Task Force; the Louisiana Wetland Protection Panel; the Coalition to Restore Coastal Louisiana; the Wetlands Conservation and Restoration Task Force; the Governor's Office of Coastal Activities Science Advisory Panel. The apparent answer to paralysis seems to have been the formation of yet another study group.

Not until 1988 did anything happen to move the plans forward. That was the year that the Coalition to Restore Coastal Louisiana, an organization which its director, Mark Davis, terms a "quasi-shadowy agent," became real. It sounds like an environmental organization, but CRCL, as it is known, is a bastard of the oil economy. Like many organizations in South Louisiana, it is driven by the collapse of the wetlands, though its raison d'être is broader. Simply put, CRCL was formed around the "vision," as Davis puts it, "that there should be a Louisiana to live in fifty years from now." That concern came not only from the state's boom or bust oil economy, but also its traditional, but only recently acknowledged, dependence on the marsh.

That knowledge came hard. In the early 1980s, oil busted. Welders and pipe fitters, roustabouts and haulers were laid off by the droves. When such downturns had occurred in earlier years, people went back to the marsh to hunt, dredge, trawl, and trap. Now they were seeing that their marsh was in tatters. The marsh could no longer support them as it had for hundreds of years.

"When you are paid $30,000 a year doing something in the oil industry," says Davis, "you don't question things. But then you suddenly have no job, and you discover that you have no natural resource. What then?" A lot of people besides fur trappers were beginning to ask that question. Local politicians, businessmen, sports fishing groups, church groups, teachers, even oil people—they were all seeing their everyday lives on the cusp of great change. "The reason CRCL works," says Davis, "is that we are a giant civics project. In a state like Louisiana, where democracy is still an experiment, we have created a vehicle for the involvement of the public and the implementation of ideas."

To Davis, "implementation of ideas," as he calls getting things done, meant knocking on the doors of influential people and putting the issue in a new light. It's not what will happen to the poor oyster-men but what will happen to everyone if nothing is done. "The cost of inaction will dwarf the cost of action," he is fond of saying.

Senator John Breaux listened to this kind of message not because he has a particular interest in the environment but because he represents Louisiana. The next step was federal passage of the Breaux Act in 1990—CWPPRA—the beginning of, and what peo-ple hoped would be, projects to stem the tide of wetlands loss. The Act guaranteed Louisiana, possessing 40 percent of the country's wetlands, an annual federal appropriation of $40 million for restoration projects. The state kicked in around $5 million a year. Given the scope of the destruction, the total amounted to table scraps from Washington.

The question was how to get this money, however paltry, to work effectively, given that so many bureaucratic agencies, from the EPA to the state Wildlife and Fisheries Department, wanted to get their hands on it for their pet projects. To galvanize action, Senator Breaux did something very impolitic: point blank, he told the disparate groups in the federal and state government, people who traditionally vied against each other for government funding, that once his bill was passed, they damn well better start working together. He also told them that they had better get some scientific expertise involved in their plans to save the coast. That's how Shea Penland became involved in CWPPRA projects. Denise Reed, a coastal geologist with expertise in freshwater marshes, who, like Penland, is on the University of New Orleans faculty, also began working with CWPPRA task force agencies. Her initial encounter with its technocrats was not pleasant. "They were suspicious of university people because they thought all we did was research. They thought I just wanted the money. I put in an enormous amount of time at enormously boring meetings to show them how science could be used. Only when you are not regarded as a prima donna in a science box do you get to be part of the team."

Such distrust defied, of course, Senator Breaux's command and has continued to compromise the effectiveness of efforts to restore the marsh. There must be a better way of thinking, a viewpoint that instills in everyone the elusive truism that halting wetlands destruction is in everyone's interest. There is. It began in the Barataria and Terrebonne estuaries. Comprised of two huge drainage and marsh systems bisected by Bayou Lafourche, the estuaries comprise many of the marshes, barrier islands, and towns already mentioned in this book. It includes Michael St. Martin's disappearing marsh, Peanut Michel's hunting grounds, Ted Falgout's offshore oil supply center, and Windell Curole's protective levees surrounding towns along Bayou Lafourche. Some 40 percent of the state's oil and gas are pumped here. Boat and ship building are leading occupations along the levees of Bayou Lafourche, employing some 15,000 people who construct a range of vessels, from nicely curved Lafitte skiffs in their backyards to powerful offshore tugs in shipyards. In many ways, the Barataria-Terrebonne Estuary, as it is known, is the heart of South Louisiana. And the 600,000 people who live here know well that their homes and livelihoods are in increasing jeopardy.

The importance of the local environment to local people struck Al Levron, a sanitation engineer for Terrebonne Parish, back in 1986. That's when sewage from an overburdened treatment plant fouled some of the marsh around Houma. A lot of oystermen were put out of work when the state closed down their polluted oyster beds. Levron grew up in the marsh, the son of fur trappers; he knew what happened when people, accustomed to its resources, could no longer harvest them. "It's like their soul disappears," he says.

He had heard that the EPA had money to put into saving estuaries through its young National Estuary Program. Estuaries, where rivers meet oceans, where freshwater mixes with salt, are undefined by a lack of definition. They are places of transition— the end of one natural system, the beginning of another. Founded on mud and sand and nurtured by decay, many lack easy majesty and have consequently lent themselves to receive the detritus of human doings.

But now their value has risen to enormous heights. They teem with hidden life, and their role as breeding ground, nursery, food source, and buffer is recognized as crucial. They are also beautiful for their muted colors, maze of channels, the mysteries they hold and suggest. In 1987, estuaries received a boost of acknowledgment when the EPA began to encourage the communities that depend on them to take responsibility for their health—water quality, wetlands preservation, wildlife, aquatic life, recreational value, and aesthetics. It was a long, idealistic, and well-intentioned list. Government was to remain in the background; the people were to organize the charge toward wise use of their prime resource, helped along by scientists, politicians, and various functionaries.

A similar effort had seen some success in Chesapeake Bay since 1983, when Maryland, Virginia, Pennsylvania, and Washington, D.C. governments, with the EPA as advisor, made a commitment to work together to save that rich estuary. Working together was the key. It meant developers listening to environmentalists and vice versa, farmers trying to understand the worries of fishermen, sewage plant managers listening to beach lovers, all for the good of Chesapeake Bay. Egos and individual interests were supposed to be set aside.

Over a decade later, the Chesapeake *is* getting cleaner. Those who depend directly on its assets, whether a developer wanting a view, or a crabber needing full traps, have managed to grasp common ground. Still, the problems are enormous, from toxic chemicals to agricultural runoff to over-harvesting of resources. Population pressure seems to override all the good intents, the multiplication of subdivision after subdivision, occupied by new faces who don't necessarily have an entrenched interest in helping the bay. Between 1970 and 1997, the number of people living around the estuary increased by 28 percent. Over 15 million now inhabit its edges.

The EPA institutionalized the ideas inherent in the Chesapeake Bay Agreement, as the negotiations surrounding efforts to save that bay came to be called, into the National Estuary Program. So far, twenty-eight estuaries are on the roster—Buzzards Bay in Massachusetts, San Francisco Bay, Columbia River estuary, Long Island Sound, and New York Harbor are some of them. The Barataria-Terrebonne Estuary was accepted into the program in 1990.

After Al Levron saw the marsh near Houma fouled by sewage, he began talking up the idea of the program. "I saw it as a way for people to sit down at the table and get some commonality going," he says. The talk went all the way to the governor's office. It was at that point, says Levron (who now directs public works for Terrebonne Parish) that "the movers and shakers got involved. I just wanted to get money for sewage control. But I like to think that I was responsible for laying the groundwork." The groundwork meant grassroots organization, getting the community involved to hammer out what the EPA demanded before it would fork over any funding—what it called a Comprehensive Conservation and Management Plan, or CCMP—essentially a mutually agreed-upon plan to save the area.

It took over one and a half years of meetings to come to an agreement about what should be addressed—freshwater diversions, pollution control, water flow improvements, etc., and education of just plain folk. But the CCMP was submitted within the EPA's deadline and the EPA did not change one word of the document.

Steve Mathies was its shepherd. He was, at the time, a young turk at the Army Corps of Engineers, a local boy who had grown up the son of a small farmer on the north side of Lake Pontchartrain. Even as a biology major at LSU, he tells me, he wanted to work for the Corps, but not as a career bureaucrat. Rather, he wanted to shift the Corps from its staunch reliance on big concrete and rock projects—environment be damned—to include the interests of oystermen and fur trappers in plans to manage nature.

When Mathies left the Corps to direct the nascent estuary program and coordinate its CCMP, he thought it would be "the easiest job in the world." He thought a plan could be drawn up, submitted to the EPA, and then they'd just wait for the check. He quickly discovered that that wasn't the way it would be. The first obstacle was one that many environmental activists stumble over: the "don't-step-on-me" factor. People tend to get jumpy when they hear that a plan is afoot that might affect the source of their livelihood, whether oil or shrimp or timber or grazing land. They tend to assume the worst. Worriers flocked to the first meeting to stop whatever they feared might be its outcome. Oil and gas people worried that the plan would force them to clean up their spoil banks; oystermen wanted to squash any proposals to install freshwater diversions for fear their reefs would be killed off; landowners didn't want anyone to dictate to them what they could, or could not, do on their land. Everyone was against something. No one was for the marsh. "I didn't know what was going on," says Mathies. "We all saw this wall we were moving toward—the loss of the estuaries—but there was no focus."

Understanding the importance of the "wall" turned out to be a blessing, indeed, a sort of goal in disguise. Everyone in Louisiana needs the marsh, though they might not realize it. This is not so apparent in the case of redwoods, or an urban park, or a wild river. And even in those first meetings to hammer out the CCMP, people didn't understand the social and historical importance of the marsh, that it was a place that had wholly supported many of their ancestors. They hadn't thought of the local beliefs and customs embedded in its muck, or that their frog and shrimp festivals depended on its rhythms.

In some of those meetings, people spoke of local history; they recounted hurricane stories and tales passed down from one generation to the next. Gradually, what had not been so obvious about the marsh became clear. The mantle of shared history descended around them. Mathies became a changed man. "I went into that job thinking that solving environmental problems is mostly hard science," says Mathies, who now works for Battelle, the huge think tank, as regional director of the Gulf of Mexico. "I came out knowing that there's as much social science involved as engineering."

Success in creating what Mathies calls "a shared vision"—the heart of the plan submitted to the EPA—came slowly. Distrust and resistance, rather than compromise and goals, blemished many meetings. Then, at one meeting, something remarkable happened—a teachable moment that should occur more frequently in discussions about the Louisiana wetlands, a lesson for all people involved complex negotiations. A Texaco official looked around the room while an environmental point was being argued and noticed that no Sierra Club members were present. The Texaco man interrupted the discussion to explain how the Sierra Club would feel about the issue—which happened to be in opposition to how he felt about it—and urged the group to incorporate both perspectives.

That moment was an epiphany.

"From that point on we all felt differently. We were ready to acknowledge that we all shared one resource," says Mathies. "People finally realized that if they came across real hard, they might be going against their brother in the oil business, or their uncle who needed a new levee in back of his house, or their cousin whose oyster reefs were threatened.

"If we don't deal with their worries, " he goes on, "we will be shooting ourselves in the foot." He leans back in his chair in his Battelle office overlooking a parking lot–littered section of downtown New Orleans. Suddenly, his face reddens with a gathering fury. "Shame on us, shame on us, for not thinking of everybody in the past." I guess that he is thinking back to his years at the Corps. I ask him about this and he says that he is. "It's this testosterone-driven business of making decisions without thinking of other people. If people feel that they are being taken advantage of, it's going to make saving the coast even more difficult."

The Barataria-Terrebonne National Estuary Program is now a major force in the state's effort to save itself. This is not to say that the two estuaries concerned stand out from the Louisiana wetlands as measures of good health. Earth-moving projects like levee removal, canal-filling, and freshwater diversions are capital-intensive and require integration with the entire coastal ecosystem.

Education has become the program's strength; its ability to bring people together to learn about and foster concern over the marsh is pervasive in a state which only recently has, as the quip goes, learned to spell the word "environment." The emphasis keeps Kerry St. Pé, the present director of the Barataria-Terrebonne National Estuary Program, on the road more than he might wish, from school assemblies to church gatherings to Corps meetings to congressional hearings, both in Baton Rouge and in Washington, D.C. His job is to find common ground and, as he says, "to invite everyone to build this house." Politicians come to cultural fairs the program holds, where cypress carvers, storytellers, and makers of handcrafted brooms talk of wetlands preservation with the same fervor as hydrologists and botanists. The politicians continue to get the respect, a problem for St. Pé. "We put these people in office that allowed the destruction. There is a tendency down here to believe that the government is taking care of the people. That has not always happened and we are trying to let the people understand the stakes if nothing is done."

The Barataria-Terrebonne National Estuary Program joins citizens together. The Coalition to Restore Coastal Louisiana uses the same tactic, quite different from the Corps's traditional bent for bulldozing aside citizen concern. Quite different from the bickering and squandering of money that the Breaux Act's CWPPRA restoration projects have entailed. There was hope here in the possibility of collaboration. Citizens began to perk up. The state government, even the Corps, even the CWPPRA agency folk, sensed that the inclusion of everyone in discussion about the vanishing coast was somehow inspiring a feeling of hope, albeit mired in frustration, rather than frustration alone.

But the marsh was still vanishing, twenty-five to thirty-five square miles of it washing away or sinking, year after year. What

would it all look like in the year 2050? The thought was frightening—another 750,000 acres or so down the drain. The phrase "system collapse" began to be bandied about. Mark Davis, Steve
Mathies, and others could see the end. They certainly saw that the
CWPPRA projects were merely Band-Aids. Davis says he sent a
letter to Senator Breaux in Washington with the warning: "Things
are broken down here." Mathies went to the top at the Army
Corps. Both of them went to the EPA. Out of these concerns was
born what Louisiana now considers its masterpiece—*Coast 2050:
Toward a Sustainable Coastal Louisiana*. Published in 1998, the 161-
page document outlined the strategies necessary to save the coast,
which were agreed upon during sixty-five public meetings that
occupied a year and a half. "Not a bad piece of work," says Davis
with obvious pride. "*Coast 2050* basically says that if you don't do
something bold, there is going to be a catastrophe." This plan was
cobbled together much as the Barataria-Terrebonne National
Estuary Program had been—by citizen input—and much along the
same lines as CRCL's emphasis on civics rather than environment.

Many people participated in its formulation—the usual gamut,
running from oil and gas company officials to alligator hunters to
levee board bureaucrats to shrimpers. They all got to have their
say, and though they might not agree in actuality, they know that
they helped develop a design—theoretically anyway—to rescue
their treasure. The strategies that they came up with are not necessarily new; they include restoring natural drainage patterns,
putting more sediment into the marsh through existing diversions,
creating marsh through dredging, plugging canals, and maintaining barrier islands. New is the effort to fit restoration efforts into
ecosystems, rather than to do a little dredging here, installing a
siphon somewhere else, and fixing up a barrier island over there.
The cost is the downside—a predicted $14 billion, a Fort Knox
amount that Louisiana knows it has to turn to Washington to get.
And that means convincing the country that Louisiana is worth the
expenditure.

• • •

The lack of money notwithstanding, the ecosystem strategy of *Coast 2050* is being plugged into projects that CWPPRA takes up. A meeting I attend, shortly after members of the Louisiana Landowners Association vented their frustration, is for the purpose of deciding which CWPPRA projects would best fit into the restoration strategies of *Coast 2050*. On the agenda for this day's meeting in February 2000 are efforts to save substantial parts of the coast: Breton Sound, the Mississippi's birdfoot delta, and the Barataria Basin—all ragged and dismal and pocked with dead cypress stands. The area, stretching from the MR. GO on the east to Bayou Lafourche on the west, includes much of the same morass that I had helicoptered over with Jack Caldwell, Sue Hawes, and the band of dazed congressional aides in the early bloom of spring 1999. Like everything in Louisiana, it is big acreage—220,100 acres of freshwater marsh; 150,000 acres of tupelo gum and cypress swamp (living and dead); 73,000 acres of intermediate marsh, meaning a little bit of salinity; 214,500 acres of brackish marsh, more saline; and 151,000 acres of salt marsh.

I had received a satellite photograph of this area, along with the meeting announcement. On the photograph, all sorts of information were superimposed, the most prominent in fine print with arrows indicating the location of different restoration projects. Some have intriguing names: Pass à Loutre sediment mining, Cutterhead Dustpan maintenance dredging, Baptiste Collette Bayou sediment diversion. Others are located only by code—MR-09, BA-05b, BA-25, and so forth. You can tell which of these are diversion projects designed to put Mississippi mud into the marsh; next to their code appear polka-dotted fan shapes indicating how the silt will spread out through the marsh. As I look them over, two things strike me: most pronounced is that given this vast area, the fan shapes are very small—most of the wetlands appear to not stand a chance of benefiting. Second is that the great majority of the names and codes refer to what are labeled in the legend as "nonconstructed CWPPRA projects." In fact, there are twenty of them, all in the minds and on the blueprints of engineers, some of them in those static states for almost a decade. Millions of dollars set aside for their construction sit in the bank. Only four CWP-PRA projects in this area have been completed. It is this kind of turtle pace that is gnawing on nerves now. *Coast 2050* is supposed to hasten salvation.

When I walk into the meeting at the Army Corps headquarters in New Orleans, the democratic principles guiding the implementation of *Coast 2050* are not fully upon me. The room is crowded with the same bureaucratic fixtures that appear at all these gatherings, plus a smattering of local color and concern—oystermen, local pols, big landowners, some CWPPRA agency guys, fishermen, planners, and a mix of quasi-green advocates keeping their eye on things. They are edgy; they want something and during their initial milling about, like minnows at current's edge, I hope what they want is for the survival of the state, not for themselves. One thing becomes certain: many have come to see whether they have a future or will have to give up their lives and relocate as the marsh disappears.

I am hearing the word "relocation" more frequently—along with "system collapse"—both uttered with quivering lips. These people's lives, or those of their constituents, are directly influenced by the fate of the marsh. As Sue Hawes scribbled out for me in the helicopter: "No marsh, no oysters." Anguish and anger gush from the audience as soon as the meeting gets under way, an airing of the scope of the crisis, an emotional release perhaps necessary for the restoration of the wetlands as the replanting of its grasses. "Plaquemines Parish is a skeleton; there's no meat on its bones." "You have no idea how bad the marsh is down there; it's goin' fast." "We in a crisis situation and somethin' got to be done 'bout the sinkin' an' all." "I'm thinkin' I got to give up my business and move north or someplace." "The coastal corrosion [*sic*] we got is so bad that if we don' do somethin' an' a hurrycane come in der, der's goin' to be a lot of dead people."

Then Windell Curole, the manager of the South Lafourche Levee District, gets up to make a statement. Curole goes to a lot of these meetings. He is a valuable spokesperson for both the environment and the state. He's an educated-looking man and wears a certain calmness. People give him their attention. Almost the first thing he told me after I met him is that he had spent a year in France on a Rotary International fellowship. Curious about the origins of his family, he tracked down some of his namesakes in Meaux (the mustard and medieval cathedral town east of Paris), whose lineage traced back to Louis XII. That he, a Cajun from South Louisiana, might be directly descended from French royalty fascinates him. He keeps rolling the idea around on his tongue as if his connection to the passage of time is too sacred a concept to merely pass off with a shrug.

Locally, he's known as the "hurricane man" down in Lafourche Parish because as manager of the South Lafourche Levee District, he knows what kind of protection is available in the form of levees when a hurricane is brewing out in the Gulf. Now, he tells it like he sees it. Though most of them have heard it before from him, they don't mind hearing it again, like a famous line from a favorite film.

The room goes quiet. "If you draw a straight line between where the state of Mississippi hits the Gulf and Texas hits the Gulf, that is our platform. South of that line to the Gulf is where we live and work. The Mississippi River gave it to us." And then, drawing an imaginary map in the air with fingers and arms, he continues. "Everything south of that line is a gift. It is our job to maintain that gift." A total hush descends. What's coming is the part people really don't like to hear. "We have to realize that some of our communities are going to have to endure sacrifices. What we have to do is decide what we want to save and what we can lose. Do we do a Korean War and draw a line in the marsh; or do we do a Vietnam War and try to save the whole thing?"

People who live along bayous south of New Orleans, where the coast is especially falling apart, feel what he is saying rather than merely understand it, as some of the planners in the room might. The reality is that land under them is disappearing, the oft-sung football field's worth every fifteen minutes. The reality is that oyster reefs are under constant threat as salt water and oyster-loving fish move north and pollutants move south. The now abundant shrimp may be scarcer in the years to come. There is profound sadness that the fur industry has been knocked out and a segment of traditional culture with it. The reality is that life in South Louisiana will never be the same again. The platform is tilting toward the Gulf and whatever is on it is beginning to roll south into the Gulf's waters.

Twenty-six of *Coast 2050*'s strategies have been earmarked for Barataria Basin alone. The agenda for this meeting is to decide which CWPPRA projects fit into these strategies. It's a rather odd request, seeing as most people don't know what goes into constructing, say, a siphon, or a sediment trap, which is a huge hole dredged out of the floor of the Mississippi into which sediment settles which is then pumped out and used to bolster the remaining marsh. And there is to be no voting at this meeting. Whatever the audience comes up with is to be by consensus. At least that's the plan. Soon, I would understand that *Coast 2050*'s philosophy and actuality follow divergent paths.

You hear the word "consensus" more these days than, say, five years ago. It's mostly associated with the Quaker philosophy where no group decision is taken short of unanimity, in short, the jury decision. People change when they understand that decisions are to be arrived at through consensus. It brings them together in a way that we do not often experience, rather than scattering them into isolated or competitive groups. It explores commonalities, rather than assuming that there are going to be winners and losers. It involves compromise, as does the democratic process, but everyone walks away with an appreciation of their would-be rival's point of view. Theoretically, anyway.

The twenty-six strategies include opening small freshwater diversions from the Mississippi River into the marsh and building barriers in bays in order to absorb wave action and thus lessen erosion. They number among them dredging projects, barrier islands restoration, and installing siphons and pumps. And each strategy, in order to pass muster, has to conform to the prerequisites of *Coast 2050* in hopes of assuring that Louisiana has a coast left by 2050.

The prerequisites are:

To sustain a coastal ecosystem with the essential functions and values of the natural ecosystem

To restore the ecosystem to the highest practicable acreage of productive and diverse wetlands, and

To accomplish this restoration through an integrated program that has multiple-use benefits; benefits not solely for wetlands, but for all the communities and resources of the coast.

They sound logical, but more importantly, they have an honest ring. *Coast 2050* is a plan that respects people *and* environment, in fact, recognizes that people are an integral part of this environment and treats both as a unit. It does not eject people from the land, close it to their use, or force them out of jobs. Environmentalists who participated in its formation had to accept the environment-people melange. Those individuals and organizations unable to accept this, who might want to create a national park out of the Louisiana wetlands, for example, found themselves out of place at the negotiation meetings.

The meeting's attendees are given handouts with a listing of projects. They are supposed to rank them "high," "medium," and "low," in terms of efficacy. The ones that receive the greatest "highs" are obviously acceptable. The participants look somewhat baffled by the procedure they are asked to participate in. Suddenly, the learning curve here becomes very steep. Suddenly, it becomes very apparent that the wonderful idea of consensus needs some adjusting to. Until the comfort zone is raised, *Coast 2050* will show itself as an unruly child rather than a disciplined student of the environment it is supposed to be saving. As I watch people fidget and murmur, I feel I know what Steve Mathies was up against during all those meetings to put together the Barataria-Terrebonne Estuary plan. Consensus is a precious concept; it takes practice just to know how to arrive at the concept.

Squabbles fizz up in about a minute. One landowner states that he doesn't want a siphon dumping silt on his land, even though he says he likes the idea of restoring the marsh. A mayor rejects a proposal for another siphon because it might compromise his town's water drainage pumps. But he, too, likes the theory. There's no budging him. Another landowner finds himself lambasted because he won't let a dredge come onto his property to dig out a canal to let freshwater in. An oysterman puts up a fuss about the freshwater diversions threatening his beds. "You want to fill up the Gulf with freshwater and send us all the way to Mexico," he complains.

Kerry St. Pé facilitates the meeting. Friendly looking, open face, ready smile, supportive type, he's good. He articulates well what is expected of the audience, stresses the theory behind consensus, explains that the whole concept of CWPPRA and *Coast 2050* is inclusive of the population and that everybody's input is of huge value.

Two hours into the meeting, things are beginning to boil over. Whispers float about that this consensus business is not working. Then murmurs course through the room like moving shadows, as the audience fumbles over how to rank projects. They don't understand what "high" versus "medium" means, and it becomes evident that no one in charge really knows, either.

"Why do we have to do this?" someone growls out.

"We all got to get together," an elderly oysterman yells out. "The Corps and the DNR is ruining this state an' ah'm ti-ared of us oystermen bein' blamed."

St. Pé is trying to hold things together. His face is getting red. He keeps asking in a louder and more desperate voice: "Do we have consensus here? Where's the consensus?" He waits. The buzz of protest intensifies. Finally, he throws his hands up. "We've got no consensus; we've got no consensus." He repeats himself a couple more times. If he had kept going, I am inclined to think he would have created a refrain for an epic poem about people in a valiant struggle to save their homes and land.

There is no formal adjournment to these meetings. Their end is a quickening death, a shuffle of a few people toward the door, then last gasps as others congregate among the folding chairs and begin little conversations, usually of reconciliation. Death has arrived when the talk turns to the perennial passions of fishing, food, politics.

By this time, I am fully aware of the paramount irony of this meeting. Unwittingly, those attending, from oystermen to Corps bureaucrats, have done the impossible; they *have come up* with a consensus. It was just the specifics that bogged them down. The big picture is as clear as could be—the consensus of the need to do what nature has always wanted to do. Of the twenty-six strategies devised to restore the wetlands, one half call for the diversion of freshwater and silt into the marsh from the Mississippi, exactly what the river did every spring before it was confined to its levees. The projects under consideration, labeled on the handout under various headings—such as diversions, siphons, restoration of natural drainage patterns, sediment "delivery" to marsh via dredges, no matter—all imitate nature. In this light, it is perhaps unfortunate that the sediment carried by the Mississippi is 70 percent less than it was a century ago due, it is theorized, to the construction of dams on tributaries upstream.

The biggest project actually restores flow—in a manner of speaking—to one of the Mississippi's former routes, now the remains of Bayou Lafourche. The sediment coursing down Bayou Lafourche over a period of three millennia built almost one million acres or so of marsh that sprawl in varying degrees of health between the Mississippi's present course and the Atchafalaya's, a huge swath of natural and commercial productivity. Here, almost three-quarters of the Gulf's commercial fish find protection and nutrients in the Barataria-Terrebonne Basin, the site of the National Estuary Program that Kerry St. Pé heads.

The delta that formed at the mouth of this former Mississippi River impeded the river's flow, forcing it to seek a new route of less resistance—its present course—to the Gulf. The process was gradual. As increasingly more of the river's current adopted its present course, its silt threw up a levee where Donaldsonville is located today. As the levee grew higher, less and less Mississippi water flowed down the Bayou Lafourche route and more and more down its new course.

When Donaldsonville first came into being in the 1750s, enough water flowed down Bayou Lafourche for the growing settlement to serve trade on both the Mississippi's new and old routes. Then towns began to spring up along Bayou Lafourche's levees, places like Napoleonville, Thibodaux, Raceland, all the way down to Port Fourchon. Though only a fraction of the Mississippi still coursed down Bayou Lafourche, it was enough to cause spring flooding in the little towns. Killing the flow was the solution to the flooding problem, completed by constructing a high levee along the Mississippi at Donaldsonville, where Bayou Lafourche began.

Only one person seemed to have objected enough to this lethal engineering to achieve a sort of immortality through resistance. His name is Walter Lemann, Sr., whose efforts, an historical marker in Donaldsonville, specifies "to restore a flow of fresh water to Bayou Lafourche after it was closed in 1903 continued to his death" eventually paid off. Now, three huge pipes cut through the Mississippi levee where the bayou used to be. This is the Walter Lemann, Sr., Pumping Station. Water pumped through them from the Mississippi churns beneath the main street and boils out a couple of hundred yards away into the bayou's channel in a peaceful thicket of willows. On the day I went there, a stand of banana plants, some grand clusters of yellow iris, a majestically frozen egret, and an abandoned tractor-trailer decorated the artificial source of Bayou Lafourche. I watched the water gush out of the pipes and head toward the Gulf of Mexico, thinking that this was hardly sufficient to nourish a million acres of marsh.

One would think a solution to the depletion of the Barataria-Terrebonne estuaries would be to bust through the artificial levee at Donaldsonville and restore the bayou, easily accomplished with minimal engineering. No surprise, this has been proposed but hooted down twice by the bayou's residents, particularly those around Donaldsonville, fearing a renewal of flooding—exactly what their marsh needs. Not to worry, a solution is in the design and talking stages. Called the Lafourche Conveyance Channel, it will be a channel to run more or less parallel to the vestigial bayou from the Mississippi. It will do exactly what nature used to do. It will cost hundreds of millions just for construction, never mind relocating homes and businesses.

Sherwood Gagliano is designing the system. He's a coastal geologist who first mapped the rate of coastal deterioration, and he has been watching the state sink for the past three decades. He's been watching the goings-on at this meeting, too, from the back of the room, and shaking his head in exasperation every once in a while. He's a small, elegantly dressed man. A lot of people, especially those who work the marshes and live in bayou towns, address him as Mr. Woody, implying a certain respect and fondness. Others, mostly academics leaning toward green, regard him differently. A former professor of coastal geology at LSU, he staked his reputation on studies he conducted that suggested that most of the state's coastal subsidence has been caused by sucking oil and gas from beneath the marsh. Gagliano's bores and seismic testing discovered whole lines of fractures in the sediment caused, theoretically, by the emptying of the support system beneath. Sections of the marsh are literally falling into the vacuum. Other coastal geologists swear that no such evidence exists.

Mr. Woody made a huge name for himself with this theory. Then, he did something that baffled his colleagues: he abandoned his pure science, resigned his academic post, and set up shop consulting for the oil and gas industry. He made a lot of money, undoubtedly fueling the bad feeling. But lately, I have been noticing him at a number of CWPPRA meetings, which seems strange because the oil industry can pay more than the government ever could. Maybe it's evidence of the demise of the onshore oil industry as pools of petroleum are nearly exhausted.

At the end of the meeting, I walk out with him into the sunshine and the glare of the Mississippi River, both of us abandoning the remains of the disjointed gathering in the gloom inside. He talks a little about the conveyance channel. He says it will carry 200,000 cubic feet per second of Mississippi water and sediment each year, building up a square mile of marsh annually. That's a lot—640 acres—by far the largest diversion project going. He tells me that it will "drive the formation of a huge ecosystem," eventually creating a new delta on the scale of the one that the Atchafalaya is building to the west.

He's a short man. He turns his head up to me, gives it a shake, and says, "We are failing miserably. The system is not working. I feel like Dr. Frankenstein. We have created a bureaucratic monster that is gobbling up a lot of money. I no longer want to be part of this destruction through my silence." I dutifully jot this down, beginning to feel like a court stenographer. He waits politely for me to finish scribbling, then starts off again. "The fallacy is that we are seeking public acceptance, but the fact is that there are only about one hundred people deeply involved. They float from meeting to meeting."

I observe that even if those in the audience aren't engineers and planners, meetings like the one we just left give them a voice and let them feel that they are part of the process.

"It's a lot of fluff," he replies. Then he stiffens and raises himself straight up. I know his message has arrived, that he wants me to carry it. "When you put up a museum, you have an architectural design competition. We need an architect terribly [himself, for example?] but we are mired in the system."

Many people think Dr. Woody may be right. But a growing number understand that the Louisiana wetlands, like any crucial ecological system, cannot be dictated to by one entity, whether it be one man or the massive Corps. It takes consensus, and though achieving consensus may be in its infancy, its goal is as democratic as the country's foundations.

Chapter Eleven

LA Goes to D.C.

The prospect of The Big One making a bull's eye on New Orleans is a summer and fall preoccupation in South Louisiana. But in late winter 2000, prospects gusted up for a big one of a very different sort. It was not coming out of the Gulf but from Washington, D.C. *This* big one would save the coast and half the state's population from *that* big one. Initially, it took the form of a huge piece of environmental legislation that would provide $45 billion over a fifteen-year period for conservation across the country. Every state would get a cut. Funds would be made available for the purchase of state and national park land and for the development of urban parks; to protect wildlife and preserve historic sites; to restore usurped lands to Native Americans; and—of enormous importance to Louisiana—to help restore coastlines. In fact, Louisiana would receive an astonishing financial package—over $300 million a year to fix the environmental disaster which the state is struggling to convince the rest of the country really exists.

In May 2000 the House passed what is known as the Conservation and Reinvestment Act (CARA) by a telling margin of 315 to 102. Substantial sentiment had accumulated for its passage, helped by election-year politics. The monies that the bill provided would go to states, sure to please voters as well as politicians campaigning for reelection, rather than being doled out by the federal government. Funding was to come from oil and gas production royalties that the federal government collected from offshore drilling. For years, these revenues had been deposited in the general treasury and used to contribute to the government's budget. Now, some of these monies would be dedicated to coastal restoration. As Louisiana produced more offshore oil and gas than any other state—resulting in substantial wear and tear of its coastline—it would receive the lion's share of the funds.

This was the dream that people in Louisiana, who are cognizant of the state's disappearance, had hoped for years would come true—a bargeful of cash from Washington. But as had happened before, the dream began to vaporize despite the whopping House vote. In midsummer, the Senate version of CARA went before the Senate Energy and Natural Resources Committee in preparation for a full vote by the Senate. After five days of debate, its members approved a far more frugal offering, a total of only $3 billion over fifteen years. Louisiana would receive a mere $175 million a year for coastal restoration—nothing to sneer at but a far cry from the $14 billion over fifteen years that the state has said it needed to prevent it from ecologically imploding. Then again, it was much more than the $40 million annual stipend provided by the Breaux Act for CWPPRA projects.

The lopsided House vote nevertheless proved that awareness of Louisiana's plight beyond its borders had taken a step forward. Louisiana Senator Mary Landrieu has lobbied ceaselessly in Washington's political corridors to foster a sense of emergency. "The time has come not to blame Louisiana but to fix Louisiana," she declared to me on the steps of the Capitol, between meetings, to promote her state. "I say to the environmental community, 'If you blame us, everyone loses.'" Then she was off in a blur of protective aides and press liaisons.

In Baton Rouge, Louisiana's capital, curiosity has developed over which of Louisiana's political power brokers will go down as the hero of the wetlands. Some people might say it's Louisiana congressmen Billy Tauzin and Chris John. Back in October 1998 they, along with

Alaska's Don Young, California's George Miller, and Michigan's John Dingell, introduced the initial version of CARA. In the Senate, Mary Landrieu and John Breaux, with Senator Frank Murkowski of Alaska and Majority Leader Trent Lott of Mississippi, presented a similar bill. In Louisiana, people refer to it as the Landrieu Bill. So perhaps she'll walk away with the credit, just as Senator John Breaux is associated with the Breaux Act and CWPPRA.

Some could say credit should go to Jack Caldwell, who revived waning interest in an early version of the Senate bill, with an emotional pitch to the Senate Energy and Natural Resources Committee. He's a good speaker in fundamentalist style, twinkly at first as though he has a joke to tell, then gradually raising his tenor, and the audience's tension, to culminate in a gush of emotion-laden sentiment. "Whole towns and communities, oil fields, coastal industries, public facilities are going into the Gulf of Mexico if we do not do something about it." Then he goes quiet. And the audience goes spiritual. Then, he's off again, building to the state's certain demise if action is not forthcoming (in the form of bountiful funding), weaving in the message that the sinking is not Louisiana's fault; it is Louisiana's victimization. Both Caldwell and Landrieu tend to wind up their arguments with a warning: imagine the cost to the nation of doing nothing.

The irony is that no one in this group is a friend of the environment. Before Caldwell became secretary of the Department of Natural Resources, he was, he proudly told me, a lawyer for the gas and oil industry, specializing in property rights. His interest in the environment is considered by many to be secondary. Landrieu, from an environmental voting record, has more right to claim leadership in saving Louisiana's environment than any other Louisiana politician. But that's not saying much. In 1999, only 22 percent of her votes were considered by the League of Conservation Voters to be pro-environment. That's pretty good for Louisiana. Her colleague, Senator John Breaux, scored zero, as did *all* of the state's representatives. Such an apparent low interest in environmental issues made a hard sell the biggest environmental package since Teddy Roosevelt.

But saving the environment for the sake of the environment was not the essential point. This bill went beyond that. Its significance lay in the fact that for the first time in this country, federal legislation that largely benefits the environment—at least the part that pertains to Louisiana's coast—had little to do with aesthetics, wildlife preservation, habitat, singularity, historical importance, etc. No, this legislation had to do with business, i.e., maintaining a healthy business environment in both Louisiana and the nation. The issue was to restore nature in order to save business, and business, in the guise of politicians and lawyers—who have little ostensible interest in the welfare of Louisiana herons or owls in the Atchafalaya Swamp—depends on saving the wetlands. Environmentalists were not necessary in this battle, thank you. National environmental groups mostly stayed out of the fray, at least in Louisiana, although they did object that passage of the bill might jump-start increased offshore drilling.[1] In March 2000, as the price of oil and gas spiraled upward, both senators Landrieu and Breaux spoke of the need for more oil drilling. Senator Breaux even declared that the time had come to open up the Arctic National Wildlife Refuge in Alaska to drilling.

Some environmental groups had warned that passage of CARA might lead some states, particularly Louisiana, to interpret coastal restoration as an excuse for highway construction. Actually, some road building is a good idea. Route 1, the mere wisp of subsiding asphalt that connects Port Fourchon to the rest of the world, is a case in point. Even an unremarkable hurricane could knock it out in hours, stranding thousands of people and compromising fuel supplies far away from Louisiana. Tropical Storm Frances, in 1998, made

[1] Though environmental groups did not weigh in in Louisiana with any particular clout during negotiations over CARA, the scene in Washington, D.C. was quite different. Under the umbrella of the International Association of Fish and Wildlife Agencies, dozens of conservation groups lobbied intensively for CARA, particularly for those sections of the bills that would restore endangered species and preserve habitat. CARA supporters also included hundreds of civic, business, and religious organizations.

Highway 90, the main artery between New Orleans and Morgan City, impassable. Until recently, sad-looking sandbags on both its shoulders remained on guard for the next flood. Updated highways in both locations could improve public safety and enhance rather than impede sheet flow of water through the marsh.

Selling CARA to Congress, from a business perspective, was a challenge. It looked so like an environmental bill, and an expensive one at that. Many people have yet to equate a healthy environment with healthy business, a lesson that Louisianans are beginning to understand. It is, in fact, *the* primary lesson that the country can learn from Louisiana. Ultimately, there is little room for separation between environment and business, between environment and people. People and the environment—so evident in the Louisiana wetlands—operate best hand in hand. But only if people treat the environment as nature treats it, rather than trying to will it to their own designs.

Selling to Congress is a ritualized, arcane art form which has, over the centuries, evolved only in nuance. It's still petitioner and court. In Washington, though, there's a special twist. You have to get an idea past "the kids" before it goes to the power brokers. Washington, in case you didn't know—and, by extension, the country—is controlled by bunches of kids, some perky with fresh master's degrees in government or public policy, some looking far too young for their bachelor's degrees.

In the spring of 2000, before the CARA bill was so enthusiastically passed by the House of Representatives, Sidney Coffee, and many other advocates for the bill, had to deal with these kids every day with great courtesy. Otherwise, the 315 to 102 vote might have been closer. In her real life, she directs communications for the Louisiana Department of Natural Resources. She controls who gets what information about the department's activities and she gets to participate in policy issues. Previously, she was the press secretary for the Delta Commission, an economic and social study of the lower Mississippi River valley, headed by former president

Bill Clinton when he was governor of Arkansas. Sidney says it was a grand experience to work with Clinton. He complimented her on good work and was her ally at press conferences when disgruntled egos made the going heavy.

Sidney likes to make things happen. She likes to work in the presence of the entitled and powerful. In the fall of 1999, CARA was stuck in Congress. Jack Caldwell asked her to move up to Washington, take an apartment there, and get the bill unstuck.

It meant she would have to leave her family for a while—her husband and two sons and her two yellow Labrador retrievers. Bereft as they were, especially, she says, the dogs, without her and she without them, she went anyway. All through the winter, she tried to dislodge the bill, spending much of her days traipsing from her office in the Hall of States Building near Union Station to the Senate and House offices.

It's a haul across the parks, the avenues, up the Capitol's steps, through endless corridors. I accompanied her on some of those hikes, curious to see how she, and by extension, Louisiana, was getting on with this potentially transformative legislation. She probably walked eight miles a day, half on concrete and half on marble, often in concert with Beth Osborne, a professional lobbyist with the Southern Governors' Association, which lobbies for issues of concern to the South. Both women are short, strong-legged, strong-hearted, and strong-lunged. Both wear their hair short, feminine, but out for business. They move through Congress as a well-rehearsed team, butting up against the kids who run our government—gatekeepers for representatives and senators. It's astonishing to realize that the information these whippersnappers accepted or rejected from these two women could profoundly influence such a crucial issue as the country's future energy supplies.

Relative good cheer abounded the day I arrived in early March 2000 in front of the Capitol. For one thing, not inconsequential to the fate of Louisiana, the temperature had shot up to 85 degrees. Daffodils nodded in tight clusters and forsythia boasted a newborn, green-tinged yellow, dazzling the city a couple of weeks ahead of schedule. Washington looked happy. No one was thinking about what this climactic anomaly—for the third year in a row—might imply for Louisiana's low-lying coast.

Producing even greater cheer among those working CARA was the apparently increasing popularity of the bill. "Three hundred seventeen—we picked up two more," was the frenzied exclamation the day I met Sidney and Beth. In other words, 317 representatives had given their nod toward the bill. As the full-house complement is 435, the bill would certainly pass, no? Not so fast. Legislation is never so straightforward. Representatives Dick Armey, the majority speaker of the house, and Tom DeLay, the whip, were apparently troubled by the bill, word in the corridors had it, so watch out. Both had the means to bring CARA's momentum to a jerking halt.

The need for a whip is a profound commentary on governance. Its etymology stems from fox hunting in Great Britain. The "whipper-in" makes sure the hounds stay in the pack. In both the British House of Commons and the American House of Representatives, the "whip" serves the same purpose—to maintain discipline among colleagues, with words rather than with whip. The power of the post, while relegated to symbolism, is still strong; if the whip is not enthusiastic toward pending legislation, the consequences for its passage can be grave.

The power of the majority speaker is more profound. The grand manager of the body, his actions propel or hold back. Simply put, a bill that is not looked upon with favor by the speaker does not appear on the agenda and therefore cannot be voted upon.

As I scurried along the halls of the Cannon Office Building behind Sidney and Beth, both deep in conversation about their imminent visit to DeLay's office, I discovered that the ritual to lobbying is nuanced. You have to find your way around the labyrinth; it's easy to get lost. Sidney and Beth were lost—and late—and walking really fast down a corridor filled with kids darting in and out of representatives' offices like bees in a hive. Petitioners cluttered the path, in slow-moving gaggles of whispered conversations. The uneasy atmosphere of seeming aimlessness is Dickensian.

"We have to get him to let this thing go to a full vote," Sidney kept saying. "How about pointing out that if he doesn't, the environmentalists will kill him," Beth suggested, facetiously, I suppose.

DeLay's office was thick with petitioners roosting on chairs like flycatchers on telephone lines, flicking their feet in nervous readiness. Their huddled shapes were discernible in frosted-glass walled cubbies behind the reception area. They bobbed in conversation toward some staff member, then arched back into isolation, the empty space quickly filling with flailing hands in ghostly dances.

Representative DeLay, or any other representative or senator, does not generally see petitioners. In fact, the chances are good that the elected official isn't even in residence, maybe back in the home district, maybe on a tour of a global hot spot. Nevertheless, the petitioners come; the office carries on; the kids listen, take notes, assess. Then they advise their boss, a tall order when experience is a classroom and maybe pamphleting during a campaign. On the Hill, these kids are called "LAs," short for legislative assistants. They're paid $17,000 to $18,000 a year, work long hours, and get to use antiquated computers that seem to be the only kind that exists in the legislative branch of the federal government. Our LA, the one assigned to hear Sidney and Beth, was Deena, a tall young woman with a long narrow face, wide blue eyes, and long sandy hair mantling her shoulders. She was very polite. The first thing she said, all apologies, was that there was no place to put us because all the cubbies were filled with other petitioners. Still uttering apologies, she gathered a few chairs in the main reception area, crowded already with two other kids flipping through databases on their cranky computers. We sat in a circle, notebooks in position. But first, the card exchange, a de rigueur ritual executed with quick solemnity at all such meetings. Deena dealt one to each of us; we gave her three. Flickering glances, as in poker, attach to top of clipboard. Someone who knows nothing about what goes on in these offices might think that our government's employees have come up with a new card game that uses miniature playing cards. They just might be right.

Meeting ready. Sidney and Beth knew they had about two minutes to deliver their message before an LA lost focus, became dis-

tracted, got impatient, or even more threatening, asked for everything in writing. When an LA asks for written material, it means
that he or she has no idea of the purpose of the visit. But courtesy
reigned, albeit clipped, the seconds ticking. Sidney rattled off the
bill's highlights—coastal restoration, wildlife conservation, land
conservation. Beth sat at attention, ready to take the baton. As a
slight downward cant of Sidney's points marked the end of her
sprint, Beth took off, bursting out of the gate with the fact that all
the Southern governors favored the bill. George W. Bush really
liked it, but he had been quiet on the issue—"under radar"—Beth
said—because if he weighed in at this point, the bill could turn into
a campaign issue, which is the last thing anyone wanted. Jeb Bush's
position, on the other hand, was more complicated. His focus at the
moment was to assure passage of pending legislation to restore the
Everglades. That entailed federal purchase of land in Florida, an
action that might risk antagonizing western congressional
Republicans worried about preserving individual property rights.
So the Bush brothers were being quiet on CARA, which did not
necessarily mean they were not in favor of it. Sidney quickly
inserted that Governor Foster of Louisiana had written both Bushes
a letter outlining what their states would receive in federal money.

Deena listened, nodding, pen poised, but didn't write a thing.
Her big blue eyes didn't seem to be registering. She was hard to
figure. Beth squirmed in her chair and then gushed out, impa-
tiently: "Well, what is DeLay's problem with this bill, anyway?"

Perhaps it was the moment Deena had been waiting for. Barely
noticeably, her face brightened and her smile broadened. Her flowing manes undulated slightly over her lifting shoulders. Then she
announced, ever so slowly, so monumentally: "Representative
DeLay is not going to stand in the way of this bill."

Sidney and Beth let out whooshes like balloons with fast leaks.
The office's tense air moved again, spiraling up to the high ceiling.
The meeting was over. Whether DeLay voted for or against the
bill was immaterial. That he wouldn't impede its introduction to
the full House was enough. The two women sashayed down the
corridors, all bubbly, trying to find their way outside.

"Maybe we've been too pessimistic about this thing," resolved Sidney, who had not been pessimistic in the slightest. "After all, it is an election year. The reps want to go home and tell everyone how they voted for this popular bill."

"Can you imagine," said Beth, "what the League of Conservation Voters would say about someone who didn't vote for the bill?"

Cheer was short-lived. They had heard that Representative Armey, the man with the power to put this bill in the closet indefinitely, was not going to be so easy. But first, lunch at Bistro Bis, a trendy work-eat place of enormous salads peppered with legislative language. The jargon coming out of Sidney and Beth, whom I can barely see over the lettuce and endives, assured me that the use of arcane words and terms is a driving force behind our government. It's all about entitlements, appropriations, dedicated funds, dedicated-funds oversights, offsets, open rules, closed rules. The two flashed the phrases around like silverware. And worry was brewing in them, too, about how to handle senators on this bill. Senators were going to need a lot of handling at this juncture because Senator Mary Landrieu of Louisiana had just introduced the new Senate version of the bill. Already, Beth and Sidney were hearing rumblings about the enormous sum of money which had been earmarked for Louisiana.

But Armey's kid was waiting. Ed was his name and, like Deena, he came right out and told us, apologetically of course, that all the cubbyholes were filled up. He hauled the chairs around in a circle in the tiny waiting area, dealt his card, got the notebook out, and he was ready. But there was no waiting for the two women's pitch. This kid was ready to pounce. Probably majored in government or political theory. In two seconds, he came right out and said that his "boss"—they're always referring to their employer as "boss" rather than "Representative"—was worried about what was known scornfully as the "land grab" portion of the bill, namely Title II. Labeled the Land and Water Conservation Fund Revitalization, Title II would do just that—revitalize the doling out of $900 million per year for the purchase of land by both federal and state governments. The fund has been considered something of a nonworking entity since it was signed into law under Lyndon Johnson. First, a lot of the authorized money was never made available. Second— especially irritating to Westerners—the government had never maintained the land it has purchased, an accusation easy to level if you don't like the idea of the feds buying up your state.

Westerners were the ones who had called Title II a "land grab," their anxiety grounded in the huge pieces of their states already owned by the federal government. Private land, as a result, was scarce, and perhaps most importantly, resentment in the states against being owned was running high. With the passage of the bill, the fear was that the Land and Water Conservation Fund— LWCF, for short—would get its act together and the government would buy up everything in sight.

While Representative Dick Armey, from the outsize Western state of Texas, may care about that, the last thing he wanted, says Ed, "is to roil the Western delegation. We don't want a floor fight over this." Further, the "boss" thought the whole debate over property rights implicit in the bill was not healthy. Sidney was prepared. She whipped out a one-pager, which she called the "Western values sheet." Its bullets spelled out how Title II would protect Western values—a euphemism for property rights—and still allow for gov- ernment purchase of land, mainly by giving states a chunk of money and the leeway to decide what land was to be purchased, thus cut- ting the feds out of the deal. It gave property owners the right to refuse offers, negotiate offers, and exchange land, too.

"This is a great Republican bill," Sidney enthused. Ed nodded, looking unconvinced.

Beth talked up the coastal restoration issue: "Can you imagine what kind of damage could be done to ports? How is food going to get to market?"

Sidney: "And keep in mind that if Congress doesn't pass this bill, then it will be probably be attached to the Omnibus Bill at the end of the year and Clinton will get all the credit."

Beth (plaintively): "I have letters from all the Southern gover- nors in favor of it."

Ed smiles and nods pleasantly. They were getting nowhere,

except giving Ed a good old time. Sensing that more good would come out of getting out of there than trying new plugs, the women rose in unison. "Please don't hesitate to use us as a resource," they echoed in tandem. "We can feed you facts." Ed looked amused, but he was not through having a little fun.

"There's another little problem with this bill," he said. Sidney and Beth sat down in unison. "It's a question of the bill's budget and entitled funds. Just a secondary concern." He shrugged. The budget at this point was a whopper—$3.021 billion per year, inclusive of all seven titles divvied up among fifty states and ten territories. California would get the most—almost $325 million, and Palau—the tiny South Pacific territory—the least—just under $570,000. The question was: if this bill passed, would there be enough money in the federal budget to pay out after the usual budgetary drains like defense, education, and health, etc., have taken their cuts. And if there was not enough in the pie, then a squabble would be inevitable.

Beth attempted to allay Ed's worry. She pointed out that this bill's budget was nothing compared to that filling the coffers of other entitled budgetary items like the Aviation Trust Fund or the ISTEA, an acronym for the Intermodal Systems Transportation Efficiency Act, which provides monies for things like light-rail construction and bicycle paths.

Ed was still not convinced. He rose from his chair with a fixed smile, arranged things on his clipboard, a sure sign that the meeting was over. Handshakes all around and polite good-byes. Sidney was just this side of ballistic when she got back to her office. Her mission had been sucked into a quagmire. Fury reddened her face. She messed up her hair by running her hands through it repeatedly. Then she got a call from someone in the Department of Natural Resources office back in Baton Rouge telling her that ABC News was doing a special on global warming and wanted to portray Louisiana as worse off than Atlantis. Her eyes sparkled with anger. "Can you imagine what those people want to do—they want to say we have already disappeared. They just want to sensationalize this whole thing. They don't want to talk about everything we are doing to stop this. They don't want to talk about CARA. They don't want to talk about what's going on right here in Washington."

She got on the phone to Louisiana, venting to me after each conversation because, I guessed, I was there. I was finding this show kind of fun, so I nodded my head a hundred times and tsked, tsked, tsked along. "And now you know what they're asking—those ABC people? They want us to take them up in a helicopter so they can film us sinking. They want us to pay for that."

A half a dozen more telephone calls, a new twist. "It's Shea Penland they've been talking to. He's been telling them that we're the next Atlantis." Shea, not one to shy away from a video camera, had told me about ABC's plans about a month earlier. But he interpreted the effort in an entirely different light. "This is going to be big," he told me. "National, international, this is going to bring some major attention to this coast."

After a few more phone calls, Sidney's face took on a mask of fixed anger. "You know what ABC wants *now*? They want Shea to take them around in a helicopter. Well, I'll be damned if DNR is going to pay to have Shea Penland fly around in a helicopter."

I suggested that she call ABC News; I knew that she had the producer's name and number. She did. She was all sweetness, explaining the fine points of CARA to the producer, and what ABC might like to put in its documentary. I gathered that ABC was all sweetness, too.

She beamed at me afterward. ABC was going to do a whole bit on CARA, she reported. They were going to interview Senator Landrieu. They were going to film some of the restoration projects I had seen from the air. And Shea Penland was going to go with them in a helicopter. DNR would be glad to foot the bill. (When ABC News aired the program in late spring of 2000, Louisiana's plight did not appear in the footage, evidence that while Washington may be concerned about the coast's recession, the national media is more focused on the fate of the eastern seaboard as sea levels rise.)

The next day, Sidney and Beth hit advocates' pay dirt in the form of Toby, a bright young South Carolinan LA in Senator Hollings's office. Sidney and Beth were beginning to go to work on senators. It was important to get Hollings early because he had "signed on," as they say, to another offshore oil-and gas-revenue-sharing bill that would provide far less money for coastal restoration. The two met in Beth's office for a pre-Hollings strategy session. Beth worked Hollings's web site to find entry points for the coming meeting. She discovered that he was the author of a bill to preserve marine life and estuaries, and that he had been instrumental in establishing a 50,000-acre bird refuge in South Carolina.

"Okay, here are some talking points," she exclaimed. "But let's ask him why he is on this other bill. There's something we don't know about this guy." She was in high-energy mode this morning, mostly, she said, because it was Friday and the LSU basketball game began at 3:15 P.M. Since her student days at LSU law school, she said, "my heartbeat goes through the roof every time I think about that team. Put that in your book," she ordered me. "That's important."

Sidney offered that she should start watching LSU basketball and noted for the record that her husband played for the LSU football team.

"Oh, yeah?" asked Beth, absorbed in the information. "What did he play?"

"Split end. He was a track star, too. You know, his 440 record was only broken four years ago." The two pondered this in admiring silence for a moment.

End of strategy meeting.

Toby looked like he hadn't slept for a couple of days. He came out of the back reaches of the Hollings complex in jeans, a battered tennis shirt, moccasins, no socks, and the beginnings of a beard. He explained that he was working on appropriations and didn't plan on getting much sleep tonight, either. The three commiserated and played the card game. Then, Sidney and Beth went at him, stressing that since his boss seemed interested in coastal preservation, he surely must be aware that the Louisiana coast is dying. Can you imagine the expense of replacing miles of railroad tracks wiped away by a hurricane, Beth mused.

Toby leaned casually against the wall of our assigned cubby, then held up his hand to stop the barrage. His problem was simple. It was the inequity of funding. South Carolina, he pointed out, exhibiting his familiarity with the bill's finer print, would get only $9 million for coastal restoration, whereas Louisiana stood to receive more than $300 million. "That's inequity," he declared.

This observation was gravy to the two women. Deftly, Sidney launched into the bill's national importance. Coastal restoration was not only a state issue; it was national transportation, national energy, national food production.

Toby looked interested.

"Protecting the Everglades is not just for Florida," Beth jumped in. "Protecting Alaska is not just for Alaskans."

"The productivity of Louisiana's wetlands is far greater than the Everglades," Sidney inserted. "They produce for the nation."

Then, in concert, both taking on facial expressions of stricken souls—a tactic I had not observed until now—they enunciated in slow motion in tandem: "We have no choice. We have got to have this money. If this money does not come through, whole communities will have to be moved."

The effect was startling. Toby was suddenly in their corner. "Do you have more information on this?" he asked. "Right now, the rationale is that Louisiana has the most drilling and gets the most money. No one has told me about the possibility of having to move whole communities."

Sidney and Beth were aglow by now, but then came the budget argument. "This bill is going to blow a hole in the federal budget," Toby predicted, but then wavered and added, "But this is a huge push for conservation and the environment. This is a good time for this bill," he announced, almost as if he were on the Senate floor himself, arguing for its passage.

It would not have done any good. In the end, Senate sentiment against what was considered by many to be a Louisiana giveaway prevailed. CARA unraveled during the summer of 2000, the massive House vote in its favor rendered meaningless. Even though sixty-three senators backed CARA, Western Republican politicians balked at what came to be consistently called a "land grab." That, and the idea that Louisiana and other states—but particularly Louisiana, with its history of squandering—should be given huge sums of money each year without any federal accountability. It was too much for some influential senators who threatened a filibuster if the bill came to a Senate vote.

Too, the Clinton administration was pushing a lands legacy bill and had abandoned its initial support for CARA. While the White House initiative shadowed some elements of CARA, it included little funding for coastal restoration. Its emphasis instead was on national monuments and historic sites, urban parks, and wildlife refuges—the Clinton legacy. Most important, Congress would be in control of the money, to be doled out as appropriations. It meant that Louisiana would have to fight for scraps each year against other states and other interests. In late spring 2000, the House voted 348 to 69 for the bill, known as the Land, Conservation, Preservation and Infrastructure Improvement Trust or, more handily, the Interior Trust Fund. Its budget, totaling $2.4 billion over a six-year period, would be devoted to conservation programs subject to appropriations approved by a rash of subcommittees. The Senate passed the bill 83 to 13 a few days after the House. At this writing, Louisiana stands to receive an appropriation of a mere $28 million, more mere bandages to stem coastal hemorrhaging.

Chapter Twelve

Coming to Terms

It takes him a while to come to the door. He shuffles over eventually, slides it open, and shoos away the dog that has been guarding the home under the great live oaks. Frank J. Ehret, Jr., eighty-three, looks okay. He is a tiny man, barely five feet high. His eyes behind thick spectacles give the appearance of being the biggest thing about him. They sparkle like stars aflame. But he says he's not well. He says he almost called to tell me not to come to visit today, but he misplaced my telephone number. His breathing has acted up. He had such chest pain last night that he was on the verge of asking his daughter to take him to the emergency room. He decided against it. He says that after three heart operations and a pacemaker and a five-month ordeal with a staph infection following one of the procedures, he'll do anything to avoid hospitals. So, I am glad that I am here just in time, to be able to meet the man people in South Louisiana call the father and founder of the Barataria Preserve, part of the Jean Lafitte National Historical Park.

The 20,000 acre park, ten miles south of New Orleans, is the only effort by the National Park Service to preserve a piece of the state's coastal environment where the public is permitted easy access and education.[1] Its facilities and educational programs are works in progress, mere shadows of the boardwalk complexes and programs offered in the Everglades. But its fauna and flora are comparable, if not more intimate. Rather large alligators reside only a few feet from some of the trails, not readily noticeable to the million-plus visitors who pad next to their waiting jaws each year. I sought Frank out because as the sole instigator for the park, he was obviously a man with a vision for the future of South Louisiana. I wondered if enforced preservation is the best way this rare land can survive.

The name Ehret is prominent in Marrero, a town on the west bank of the Mississippi where the preserve is located, a thirty-minute drive from downtown New Orleans. There's a high school named after Frank's grandfather, part of the Jefferson Parish school system, where Frank spent his professional career as a special education teacher. Frank lives on Ehret Street. His ancestors came here from Germany in the 1720s, when the nascent French colony was in the throes of some severe growing pains. New Orleans was filling with settlers, mostly craftsmen and vagabonds. They knew little about growing crops. The Indians who had helped supply the newcomers with food could not keep up with the demand. Starvation was approaching. Germany came to the rescue, sending thousands of small farmers who settled the west bank of the Mississippi and raised New Orleans's table food. The Ehret family settled back then in just about the same spot as where Frank now dwells.

When Frank was a young man, Marrero was a river town, with a main street, neighborhood bars, dance halls, and small-town intimacies. It had changed its name from Amesville, founded around the docking facilities that the Texas Oil Company, the forerunner of Texaco, built on the river. Oil from Spindletop Field in Beaumont, Texas, was barged up the Mississippi to the docks and sold to the sugar plantations in the area. During the annual late-fall processing of cane, the unrefined oil was used as fuel to boil down the syrup.

[1] While the National Park Service does maintain other facilities and sites in Louisiana, the Jean Lafitte National Historic Park is its only extensive holding.

Now Marrero has lost its connection to the river. Route 90 bisects it in one direction, Barataria Boulevard in the other. Enormous shopping plazas collide with each other, rendering a boxy, neon-lit landscape of Piggly-Wigglies, Home Depots, Wal-Marts, and the occasional pawnshop. And then the subdivisions begin, marching in step formation into the marsh.

Keep going down Barataria Boulevard past the subdivisions, and you will come to Frank's life achievement, *his* park, really, of swamp and marsh, bayous and great cypress trees. It took him a quarter century of lonely battle to save this land from the developers. Now part of the National Park Service, it could be seen as an answer to South Louisiana's crisis—protected and preserved forever, open to visitors but not to dwellers.

Frank is proud of his triumph, as he should be. Most people have forgotten that he is largely responsible for the park. He likes to tell the story about its founding. Besides, his wife has died; his grown children are gone, though they live nearby with their own families. The house is too quiet. So, with grace, he sweeps me into his living room where the TV is blaring to an audience of one whose mind is much too lively for its fodder.

The dark-paneled walls of his living room are covered with plaques and awards from allies-come-lately honoring him for his long fight. The most recent commendation is dated 1989, the year he received the National Conservation Achievement Award from the National Wildlife Federation. The plaque recognizes him as a hero, a man who grabbed a chunk of nature and set it in stone like a diamond, away from the grasp of greedy neighbors who wished to reap financial rewards from it, as have the oil and gas companies and land grabbers like Edward Wisner for generations.

The walls also hold evidence of another side of Frank's life—his life in the swamps where he, ironically, appears to have held little sacred. Photographs of Frank and his friends holding up strings of fish, of shot deer lying at their feet, of piles of ducks, testaments to a life of abundance that the marsh has always provided are lined up like trophies. On a mantel, exquisitely detailed and painted decoys collect dust, carved by Frank out of cypress or tupelo gum, some used in the hunt. A couple of half-finished ones clutter the threshold. They will never be completed, Frank tells me. His eyes have given out on him for detail work.

The photographs and decoys say more to me than all the framed accolades. They speak of a place in nature that is recklessly pillaged rather than treated with the respect due a nurturer and provider over the centuries. The absurd bounty of bleeding wildlife complicates for me the picture that I am trying to bring to focus of Frank, the preservationist as well as the hunter. Is he any different from those many people in South Louisiana who tolerate the dredging of canals haphazardly through the wetlands and the dumping of pollutants on their fragility, and then profess such love for this land, all in a bewildering mix of adoration and abuse?

He begins to talk, fumbling through a battered folder. I have to help him, and what comes out is a large-format scrapbook he has pasted together of the volumes of newspaper articles over the years about his victories and defeats and, at last, his ultimate triumph. Finally, he turns the TV off and offers me a seat. He is out of breath and I urge him to slow down. He shrugs off my concern, insists on standing, and talks terribly fast.

He tells me that what has happened to his beloved South Louisiana—its disintegration and disappearance—is an affront against nature. As he tells me how he has seen and felt the land die, I begin to equate him with a sort of prodigal son. A young man now become an old man, he once may not have understood that these wetlands had any end to them, but then realized that they, like the deer and ducks he hunted, can be killed. He came back to the land to embrace and protect it.

As a boy, he hunted and fished Barataria Bay with his daddy. His daddy loved the land, too; he ran a charter boat service between Lafitte and Grand Isle, for weekenders going to the Oleana Hotel, still in business near the beach behind the levee that holds back the rising Gulf. A photograph of the boat on the living room wall reveals her to be a beauty, a forty-five-foot converted oyster lugger named the *Snappy Stepper*. She had a sleek cabin running her length and a full deck overhead where the Kid Thomas

Jazz Band played as the boat plied Barataria Bayou. When it neared Manila Village, all the Chinese shrimp fishermen and dryers would come out on their skiffs and start dancing on the aft decks. Oh, it was a sight, says Frank, his eyes burning.

In the 1950s, the developers came. They weren't strangers from the Midwest. Frank had gone to school with them, fished and hunted with some of them. These men had looked across the river to New Orleans which was beginning to sprawl. They saw the marsh in their backyards not as a continuation of their lives but as the beginning of new lives. They saw money and began to talk about the beautiful homes that would soon line the bayous and drainage canals. Frank saw a disaster in the making. There was Huey Long on a stump, saying that this was the best thing that had ever happened to Louisiana. He said he was going to sell all the land around Barataria Bay for a dollar an acre and make the state rich and the people happy. He commended the developers as heroes of the state. He said they were turning wasteland into real land.

Frank already knew that the marsh was real enough. He also knew it was sinking and that any houses that went up on it would begin settling like a heavy book on a cushion. The money and greed were bad enough; what also angered him was the deception of wringing the marsh out like a sponge.

In the early 1960s, Frank began talking up the idea of preservation with the same fervor that the developers were exuding in their glorification of new subdivisions. Frank was alone; the developers had money and friends. But Frank did have an infectious spirit; he got himself elected to a minor political office, a space to give himself a platform. "I created my own road show to get this park idea off the ground," he tells me as he flips through his scrapbook. He spoke at schools, he wooed the local press, he gave speeches wherever he could. Poking around the would-be park, he discovered a couple of old cypress *pirogues* submerged in a bayou along with arrowheads and pottery shards. He wielded them like courtroom evidence to demonstrate to the public what would be lost if the land were allowed to succumb to dredge and bulldozer—its history, its spirit gone up in lawns and concrete slabs.

Even after he tried for years to sell the park idea, people still came up to him and wondered aloud, "Who the hell would want to visit a swamp?" The real estate developers just kept on gouging drainage canals out of the marsh. "I take defeat as an asset," Frank tells me, his voice going hoarse and his breathing, shallow. "I never give up."

His persistence paid off. In 1966, Governor John McKeithen designated the area a state park, a victory to be sure, but hollow as it turned out, for the state claimed it had no funds to administer the new holding. The publicity led former Senator Bennett Johnston to take notice, though. Frank, with Johnston's advice, wrote a proposal to create a national park. Johnston shepherded it through Congress, and in 1978, President Jimmy Carter signed the legislation into law. Another photograph on Frank's wall shows him with President Carter, the bill in front of them, Frank beaming.

I visited the park on a spring day when fields of blue irises bloomed among the cypress and spider lilies rose from pockets of black water like fairy wands. The trail I chose to hike ever so gradually descends through the remains of a cypress forest, into a brackish marsh and finally to the banks of Bayou Segnette, which connects to the Gulf. From the bayou's banks, you look west across Lake Salvador and the marshes on the opposing banks, and you can think that there is no other place so rankly primeval as right here. The bustling serenity of the exuberant growth around me struck me as the way this land was meant to be. I could sense the flow of water through the swamp, too slow to see, but feeding and filtering on its course to the sea. I could see the demarcation between swamp—where the cypress grow—and freshwater marsh with its profusion of flora, and then brackish marsh where cordgrasses predominate.

Naively, I thought that the whole coast of Louisiana should be included in this park. From time to time, proposals have been floated to try to preserve a large part of the coast along the same lines as the Jean Lafitte National Historical Park's Barataria

Preserve. They have all been short-lived, as they should be. More than the Everglades, South Louisiana has always been a working coast, prolific in natural resources, industry, and local culture. The combination of rich nature and rich human bustle is incomparable in this country. Here is South Louisiana's possibility: the melding of a working relationship between humans and nature. Developing that balance is the struggle that South Louisiana is engaged in right now. The heritage of this land is its oil men and shrimpers, fur trappers and oystermen, as much as its cypress stands and lost lakes in the marsh. Pure nature by itself, grand as it may be, is illusory in a place where human beings have for so long been implanted on the land.

Just a few miles from the preserve, the little town of Lafitte lines up its buildings along Barataria Bayou—oyster plants, fish processors, tilting houses, a couple of gas stations and delis—all pretty traditional for a bayou town, except for the recent additions of a gift shop and a bed and breakfast. The road ends just south of the town. The marsh takes over from there, and flows all the way to the Gulf, though its wholeness is severely interrupted by the crisscrossing canals of the Lafitte oil and gas fields.

The contrasting juxtaposition of the preserve and the working town nicely exemplifies the ethical impossibility of South Louisiana putting itself into a state of preservation. As lovely and peaceful as the preserve may be, its nature is compromised by the lack of human activity within its borders. On the other hand, the chewing up that the marsh has endured a few miles away under the auspices of the oil and gas fields speaks to the need for controls.

A stable relationship between people and nature in South Louisiana is still far from a reality. Movement in that direction is a slow crawl, a tedious one, as the meetings stretch on, day after day. And as the wetlands inexorably decline, day after day, the fear is palpable. Exciting to witness, however, is the growing realization among people from all walks of life along the Gulf that if they do not conduct themselves on nature's terms, all will be lost. The natural system, pushed to the limits, is rebelling. The marsh will continue to sink, the grasses die as salt water encroaches, the marine bounty vanish, and the oil and gas production remain under threat of upheaval, if compromise does not outweigh dominance as a way of life.

Compromise means that enormous quantities of water have to be allowed to flow into the wetlands from the Mississippi River, despite its diminished silt load. That's the emerging consensus, not always willingly acknowledged but in everyone's mind. Let the water flow. Let the nutrients and silt replenish the sinking land. It will cost billions of dollars, but that's what nature demands. That's what nature got before the levees and the canals went in.

The first European settlers to this country were daunted by the interminable forests before them. Remnants of unease may still run in us. Of the Pilgrims trying to survive their first winter in this country in 1621, William Bradford, Pilgrim elder and governor of Plymouth, recorded in his journal, "They that know the winters of that country know them to be sharp and violent, and subject to cruel and fierce storms. Besides, what could they see but a hideous and desolate wilderness, full of wild beasts and wild men?" The solution was to "civilize" the new land. "Gardens may be made [out of the wilderness] without expense," Thomas Jefferson wrote a century and a half later. "We have only to cut out the superabundant plants."

The debate between preservation on one hand and conservation, or "wise use," on the other, has played back and forth across the land over the generations. John Muir, the archetypal preservationist, who founded the Sierra Club and, indirectly, the National Park Service, believed that nature was God's temple which intemperate human beings had no right to alter. "The forests of America, however slighted by man, must have been a great delight to God," he wrote, "for they were the best he ever planted. The whole continent was a garden, and from the beginning it seemed to be favored above all the other wild parks and gardens of the globe."

Gifford Pinchot, founder of the U.S. Forest Service, disagreed. He argued for "wise use" of natural resources, control, and management for the benefit of all. "The first principle of conservation is development, the use of the natural resources now existing on this continent for the benefit of the people who live here now," he declared.

The debate continues. In 1980, then-president Jimmy Carter signed into law the Alaska National Interest Lands Conservation Act, which would preserve over 100 million acres—almost one-third of the state—into ten new national parks, wildlife refuges, and wilderness areas. Alaskans, prodded by big oil money and big politics, were outraged. They took out their anger on a Carter effigy and hurled bottles at it. Now, they consider Carter a hero, as receipts from tourism have topped one billion dollars a year. Even so, sentiment runs high against efforts to give national-monument status to the 20-million-acre Arctic National Wildlife Refuge on Alaska's north coast. The oil and business interests are lobbying hard for the whole 20-million-acre expanse to be opened to oil drilling.

This country, so abundant in resources far beyond the Louisiana marsh, has never shown a talent for restraint in using them. Forests, fish stocks, water resources, topsoil, Western grazing lands, clean air, and genetic diversity—these come to mind, for starters, as natural gifts which now run scarce from overuse and lack of stewardship. But ask almost anyone if they are aware of shortages of anything in this country and it's likely they'll say "no." It's almost dead sure that urban folk don't know the meaning of being without, provided that supermarket shelves remain stocked. Behind the abundance, we are just mustering through; scarcity may be coming down the road.

Take groundwater. Not many people think about it. Half the country gets its drinking water from wells rather than from municipal water systems, which are apt to filter water. The United States Geological Survey (USGS), in a survey of shallow wells across the country, found that 15 percent are high in nitrates, which can cause the potentially fatal "blue-baby" syndrome. In another survey, the USGS found sixty volatile organic compounds in both urban and rural wells. Leading in the pollutants was methyl-butyl ether (MTBE), a compound formerly added to gasoline for the purpose, ironically, of reducing carbon monoxide in exhaust, another pollutant. Still another sampling came up with two or more pesticides in well water. A certain disregard has led to this worsening situation. As links between introduced chemical compounds in water and cancers and infertility become more plausible, both concern and research are increasing. Somehow, the obvious tends to get missed: stuff that is put onto or into the ground is bound to end up in the water beneath it. The too-easy assumption all along is that these substances, though perhaps invisible, will just go away. We can do what we want with the land; no one will be worse off, is the assumption.

Or fish stocks. The sardine fishery off Southern California collapsed some years ago; the cod fishery in the North Atlantic crashed more recently and is just showing signs of coming back; bluefin tuna are so rare that Japanese restaurateurs are willing to pay obscene sums for them and transport them from a Long Island dock to Tokyo by charter jet for sushi that same evening. One 715-pound giant sold for $83,500, or $117 per pound, to be divided up into 2,400 servings at $75 each for a gross profit of almost $100,000.[2]

Swordfish, the backyard barbecue delight, may be headed toward commercial extinction due to mismanagement of the Atlantic population since the advent of long-lining in the 1960s. The big swordfish, the 200- to 600-pounders, are long gone. Females do not reach sexual maturity until they grow beyond ninety pounds. Over half the swordfish taken in recent years, however, have been immature. And the legal size limit is a mere forty-four pounds. With every small swordfish taken, fishermen are seeing their livelihood disappear.[3]

Marketing strategists have come up with a solution: get the American public used to alternative species—sharks, marlin, orange roughy, Chilean sea bass, to name a few. Now, many shark species are

[2] These figures were provided by Carl Safina, director of the National Audubon Society's Living Oceans Program.

[3] Recent legislation has imposed seasonal moratoriums to long-line fishermen on swordfishing in nursery grounds covering over 130,000 square miles of the Atlantic. It is expected to result in a reduction of juvenile swordfish catches of between 31 percent and 42 percent.

in trouble. Marlin are overfished. Orange roughy and Chilean sea bass, both deepwater species that mature extremely slowly, are nearing depletion.

The will to acknowledge limits appears to be lacking. But a few telling exceptions exist. Cod fishermen in the little outports that dot the rocky coast of Newfoundland, Canada, are one exception. In the 1980s, they warned that stocks were dwindling. Though cod fishing was their only income source, as it had been for their fathers and grandfathers, many volunteered to reduce their catches in hopes of reviving the species for future generations. The effort fell flat, however, as international pressure mounted to keep the offshore fishery open. The result was a severe crash, making the North Atlantic cod fishery one of the more notorious examples of poor management.

The attitude of the small fishermen is an interesting exception to the rule, one that may well have sprung out of familiarity with a resource. Such intimacy is rare today, as managers from afar determine the health of populations, whether fish or timber, and try to fulfill the demand. Late in 2000, George Barisich told me that he had been forced to lay off shrimping for a while, until "the jellies" pass through. He said he had never seen such big ones. "They just clog your nets up like big balloons." George did not know why the jellyfish were passing through his trawling grounds in such numbers.[4] He wasn't too bothered, though. He said that he had plenty of oysters to dredge in the meantime. "What with the lack of rain, they're real salty, just as good as they get," he happily reported.

[4] The burgeoning population of moon jellyfish, a species native to the Gulf of Mexico, is becoming an increasing threat both to the region's finfish species as well as to shrimp. Changes in the environment are responsible. First, jellyfish require hard surfaces for spawning, and the increase in offshore oil platforms well serves this need. Second, jellyfish—filter feeders—are thriving on the Gulf's rich assortment of plankton and bacteria, a result of nitrogen pollution carried by the Mississippi, also responsible for the "dead zone." The result is that jellyfish are sweeping inland waters clean of larvae and eggs of shrimp and finfish. An exacerbating problem is the recent appearance of the Australian spotted jellyfish, a basketball-sized blob that jumped from the Pacific to the Caribbean and is slowly drifting northward.

Peanut Michel was happy, too. He had just heard from the Louisiana Department of Wildlife and Fisheries that his alligator quota for the fall season had been increased from sixty to seventy-five. And even better, he had been lucky enough to be a winner in an "alligator lottery" which would permit him and fourteen other hunters to kill 1,500 alligators on Marsh Island. The reason for Peanut's good fortune had to do with the wise management of the state's alligator population, so robustly revived since overhunting almost did the creatures in.

The management did not come from Washington or Baton Rouge. It came from field researchers who grew up in the marsh and cruised the same bayous as alligators. Such familiarity has allowed them to feel the reptiles' presence, know when they're around, and when they're not, know when something might be amiss.

The debate between preservation and wise use continues, but perhaps it doesn't have to, judging from the dilemma facing Louisiana. The ability to sense the land was Frank Ehret's gift. I have always been struck by the passion that South Louisianans express for their corner of the country, far more so than elsewhere. Their exuberance shows in different ways; Frank's was one way, unique and powerful. A biologist for the Army Corps of Engineers showed me another way, as did some birdwatchers on Grand Isle. The Corps biologist and I were casting for redfish and speckled trout and chatting about the coast's plight just off the Chandeleurs—the beautiful arc of barrier islands, or what's left of them, toward the Mississippi State line. "I grew up taking as much as I want," he said. "My daddy took me duck hunting when I was a kid and we shot so many we didn't know what to do with them. We have so much abundance here we couldn't conceive of not keeping what we shoot or catch."

I had heard other people in South Louisiana talk about the wetlands' seemingly endless bounty. Then the biologist added another perspective. He said that he had traveled out of South Louisiana only a few times. Each time, he couldn't wait to get back. He hoped never to have to leave. "I just don't have any interest in anyplace else. I believe right here must be the best place in the world. Maybe we are going to have to start changing the way we think about all this abundance if we want to keep it that way."

Louisiana's bird-watchers show a gentler side of an appreciation for nature. From April to June, the people of Grand Isle enjoy a glorious migration of neotropical birds. They arrive exhausted after their four-hundred-mile flight across the Gulf of Mexico to recuperate in the few remaining live oak thickets. On an April day that I went birding with a handful of people from assorted parts of the state, led by the Nature Conservancy, scarlet tanagers had arrived in all their showiness along with seven or eight warbler species. We did not have to look hard for them. They were resting up, actually, in a neighborhood of mobile homes and dog-patrolled bungalows shaded by a canopy of live oaks. Residents emerged from screened porches to point out different species while kids on noisy off-road vehicles crashed through the trees on thrillingly rutted paths. The birds didn't seem to mind. One man came trotting toward us pushing a pram with a whining child in it to announce the sighting of a black-throated green warbler. He was very excited about his finding and hushed the child as he pointed out the direction. We set off and soon came to a trailer with a little clearing in front in which the owner had set up a huge bird feeder. A grand display of indigo buntings and purple grosbeaks were feeding, a sight that melted some of the purveyors. Before we left, the owner of the feeder, a retired oysterman who had allied himself with birders, made everyone sign his book. He said he was going to "take it to Washington and show those congressmen how much we care about our state down here." With birding turning into one of the most popular pastimes in this country, surpassing walking, hiking, and other outdoor recreations, it's too bad that the state does not market its living natural riches.

I keep thinking of Frank Ehret and his will to fight for the land. I keep thinking of the attachment to the wetlands that grows in people who have moved to South Louisiana from other places. And those who were born in the marshes rarely leave them; if they do, they tend to return, with a sigh of relief. Families are huge here, whole towns occupied by people with three or four last names— Arceneaux, Babineaux, Charamie, Curole, Landry, Petrie, Stelly, Voisin—like the beginning of a child's song. I think of the efforts to come to consensus to stop the coast from dying. And I think of the water lapping under the stilts that hold up bayou towns, and the worry on faces up and down the bayous. In the way people here regard their coast, with a mix of religious obsession and practical need, I see salvation. Few other places rich in nature in this country are just as rich in human activity.

That's one reality. There's another reality, or rather a possibility, that is just beginning to emerge in South Louisiana's quandary over its environment. Something that Joe Suhayda, the computer simulation hurricane forecaster, said comes back to me, about the people around here needing to know what they want to do. They have all the ingredients to set things right with the environment and still live in the environment that they have created. They have the spirit, the love of land, the desire to involve everyone in decisions, the money, but they don't have the plan. Not yet. They could learn something from those Newfoundland cod fishermen.

At all the meetings I attended during a year and a half of research, meetings at Corps headquarters, with representatives of the Barataria-Terrebonne National Estuary Program, at universities, with oystermen and shrimpers, or with oil men, I found myself squirming with excitement in my seat as I watched this slow-motion struggle to find The Plan. "They're almost there," I kept telling myself. "They're finding their way." Then I realized they hadn't, that their direction had become mired in some bureaucrat's meanderings, some oysterman's complaints that his reefs would be buried by all the silt let in from the Mississippi, or in just plain indecisiveness. But when the next meeting came along, I went full of expectation that this time it was going to happen.

The pressure is mounting to save the coast. These days, you can feel the tension rise more quickly in meetings. People know that an end is coming, whether by hurricane, by encroaching saltwater, by the rapidly rising sea itself, or just by the salt grasses giving up and dying as seemed to be happening during the summer of 2000. They know that the cornucopia is emptying. They also know that they all make their livelihoods here and that while some of them work the oil rigs, their uncles and cousins work the oyster reefs. The wetlands have brought these disparate groups together. It is their bed and home and the entire country's bounty, not as a park but as a working arena of life. No one can give that up and remain true to the knowledge that we are all part of nature.

Over three and a half million acres of wetlands still remain in Louisiana. Their future remains in doubt. *(Philip Gould)*

Selected Sources

The information for much of this book was garnered through scores of interviews, field trips with various scientists, attending sequences of meetings, and just by being present in this unique part of the country. The following sources, however, substantially contributed information to various chapters.

Chapter 2

Boesch, Donald F., et al. "Scientific Assessment of Coastal Wetlands Loss, Restoration and Management in Louisiana." *Journal of Coastal Research*, Special issue No. 20 (May, 1994).

Gagliano, Sherwood, and J. L. Van Beek. *Geologic and Geomorphic Processes of Deltaic Processes, Mississippi Delta System.* Center for Wetland Resources. Hydrologic and Geologic Studies of Coastal Louisiana. Report No. 1. Baton Rouge: Louisiana State University, 1970.

Kesel, Richard H. "The Decline in the Suspended Load of the Lower Mississippi River and Its Influence on Adjacent Wetlands." *Environmental Geology and Water Sciences*, Volume 11, No. 3 (1988): 271–81.

No Time to Lose: Facing the Future of Louisiana and the Crisis of Coastal Land Loss. Coalition to Restore Coastal Louisiana. (Revised, 2000.)

Turner, R. E. "Low-Cost Wetland Restoration and Creation Projects for Coastal Louisiana." In *Natural System Function and Response to Human Influence, A Symposium 1998*. Baton Rouge, Louisiana: Sea Grant College Program, 1999.

Vairin, B.A. *Caring for Coastal Wetlands: The Coastal Wetlands Planning, Protection and Restoration Act.* Lafayette, Louisiana: U.S. Geological Survey. National Wetlands Research Center, 1997.

Williams, S. Jeffress, Shea Penland, et al. (eds.). *Atlas of Shoreline Changes in Louisiana from 1853 to 1989.* Washington, D.C.: U.S. Geological Survey, 1992.

Chapter 3

Burkett, V. and J. Kusler. "Climate Change Potential Impacts and Interactions in Wetlands of the United States." *Journal of the American Water Resources Association*, Vol. 26. No. 2. (2000): 313–19.

Gosselink, J.G. *The Ecology of Delta Marshes of Coastal Louisiana: A Community Profile.* Baton Rouge: Louisiana State University Press, 1984.

Komar, Paul D. "Shore Leave." *The Sciences*, Volume 40. No. 1 (Jan.–Feb. 2000): 20–24.

Penland, Shea. *Barrier Island Evolution, Delta Plain Development, and Chenier Plain Formation in Louisiana* (Ph.D. diss.). Louisiana State University, 1990.

Williams, S. Jeffress, Kurt Dodd, and Kathleen Krafft Gohn. *Coasts in Crisis.* U. S. Department of the Interior. U.S. Geological Survey. U.S. Geological Survey Circular 1075. Washington, D.C.: 1990.

Chapter 4

Baker, Brian J., John M. Englhardt, and James D. Reid. "Large Scale/NORM/NOW Disposal through Slurry Waste Injection: Data Analysis and Modeling." Presentation before the Society of Petroleum Engineers Annual Technical Conference and Exhibition. Houston, Texas, October 3–6, 1999.

Costanza, Robert, et al. "The Value of the World Ecosystem Services and Natural Capital." *Nature*, Vol. 387 (May15, 1997): 253–60.

"Fourchon Reporting Update." Report prepared by Chevron Pipe Line Company for the U.S. Coast Guard and the Louisiana Department of Environmental Quality, 1993.

Hallowell, Christopher. *People of the Bayou.* New York: E.P. Dutton, 1979.

Konigsmark, Anne Rochell. "Swamp Folk: Cajun Lawyer Sounds 'War Cry' as He Battles Oil Companies to Save the Marsh." *The Atlanta Journal and Constitution* (February 21, 1999).

"Louisiana Energy Facts Annual, 1998." Department of Natural Resources. Technology Assessment Division. Baton Rouge, Louisiana, 1999.

Lindstedt, Diane M., Lori L. Nunn, et al. *History of Oil and Gas Development in Coastal Louisiana*. Baton Rouge: Louisiana Geological Survey, 1991.

Natural Recovery and NORM Constituent Monitoring Plan. Report prepared for Chevron Research and Technology Company by Ecology and Environment, Inc., 1999.

Odum, Eugene P. *Ecology and Our Endangered Life Support Systems*. Sunderland, Mass.: Sinauer Associates, 1989.

Padgett, H. R. "The Marine Shell Fisheries of Louisiana" (Ph.D. thesis). Louisiana State University, 1960.

United States District Court, Eastern District of Louisiana. *Partial Findings and Conclusions. Michael X. St. Martin and Virginia Rayne St. Martin vs. Mobil Exploration and Producing U.S., et al*. Civil Action No. 95-4128. August 13, 1998.

United States District Court for the Eastern District of Louisiana, Appeal. *Original Brief for Michael X. St. Martin and Virginia Rayne St. Martin*. Civil Action No. 95-04128.

Schleifstein, Mark. "Lawyer Blames Erosion on Oil Firms. Two School Boards Are Recent Clients." *The Times-Picayune* (April 12, 1999): p. 1B.

Turner, R. E. "Wetland Loss in the Northern Gulf of Mexico: Multiple Working Hypotheses." *Estuaries*, Vol. 20, No. 1, March 1997.

Chapter 5

Beardsley, Tim. "Death in the Deep." *Scientific American*, 277, No. 5 (November 1997): 19–20.

Barisich, George. "The Government Went Down to Louisiana: A Song Written for a Cause." Unpublished.

Dortch, Q., et al. "What Is the Threat of Harmful Algal Blooms in Louisiana Coastal Waters?" *Natural System Function and Response to Human Influence, A Symposium 1998*. Baton Rouge: Louisiana Sea Grant College, 1999.

Greenberg, Paul A. "Fisher Spearheads Crusade to Fight Industry Setbacks." *The Times-Picayune* (April 6, 1997): pp. G36–39.

Margavio, Anthony V. and Craig J. Forsyth. *Caught in the Net: The Conflict Between Shrimpers and Conservationists*. College Station, TX: Texas A & M University Press, 1996.

Rabelais, Nancy N. "Harmful Algal Bloom Research and Control Act of 1997 with Specific Comments on Hypoxia and Nutrient Enrichment." Testimony to the U.S. Senate Subcommittee on Oceans and Fisheries. May 20, 1998.

Rabelais, Nancy N., Eugene R. Turner, et al. "Characterization of Hypoxia: Topic 1, Report for the Integrated Assessment on Hypoxia in the Gulf of Mexico." U.S. Department of Commerce. National Oceanic and Atmospheric Administration Coastal Ocean Program. Decision Analysis Series No. 15 (May 1999).

Chapter 8

Brasseaux, Carl A. ed. *A Comparative View of French Louisiana, 1699 and 1762: The Journals of Pierre Le Moyne d'Iberville and Jean-Jacques-Blaise d'Abbadie.* Lafayette, LA: University of Southwestern Louisiana. Center for Louisiana Studies, 1979.

Chase, John. *Frenchmen, Desire, Good Children and Other Streets of New Orleans.* New Orleans: Robert L. Crager & Co., 1949.

Creté, Liliane. *Daily Life in Louisiana, 1815–1830.* Translated by Patrick Gregory. Baton Rouge, LA: Louisiana University Press, 1981.

DuFour, Robert. L. *Ten Flags in the Wind.* New York: Harper & Row, 1967.

Du Pratz, and M. Le Page. *The History of Louisiana.* Translated by Joseph G. Tregle, Jr. Baton Rouge, LA: Louisiana State University, 1975.

Johnson, Jerah. "Colonial New Orleans: A Fragment of the Eighteenth-Century French Ethos." in Arnold R. Hirsch and Joseph Logsdon. *Creole New Orleans: Race and Americanization.* Baton Rouge, LA: Louisiana State University, 1992.

McWilliams, Richebourg Gaillard (ed. and trans.). *Fleur de Lys and Calumet, Being the Penicaut Narrative of French Adventure in Louisiana.* Baton Rouge: Louisiana State University Press, 1953.

McWilliams, Richebourg Gaillard (ed. and trans.). *Iberville's Gulf Journals.* Tuscaloosa, Alabama: The University of Alabama Press, 1981.

Plaisance, E. Charles. *The Hurricane of 1893 at Cheniere.* Galliano, Louisiana: Harris J. Cheramie, 1981.

New Orleans, LA, Sewage and Water Board. "Report on Hurricane Betsy, September 9–10, 1965." October 8, 1965.

Chapter 9

Allen, Scott. "Water Futures." *Boston Sunday Globe* (November 7, 1999): p. A1.

Cowdrey, Albert E. "Land's End: A History of the New Orleans District, U.S. Army Corps of Engineers, and Its Lifelong Battle with the Lower

Mississippi and Other Rivers Wending Their Way to the Sea." Washington, D.C.: U.S. Army Corps of Engineers, 1977.

Dunne, Mike. "Coast in Peril." *The Advocate*. Part of series (11/7/99 to 11/11/99).

Chapter 10

"Coast 2050: Towards a Sustainable Coastal Louisiana." Report of the Louisiana Coastal Wetlands Conservation and Restoration Task Force and the Wetlands Conservation and Restoration Authority. Baton Rouge, LA: Louisiana Department of Natural Resources, 1998.

Mathies, Steve. "Implementation of *Coast 2050*." Unpublished paper.

Moore, David M. and Robert D. Rivers. *The Estuary Compact: A Public-Private Promise to Work Together to Save the Barataria and Terrebonne Basins*. Comprehensive Conservation and Management Plan—Part 2. Thibodaux, LA: Barataria-Terrebonne National Estuary Program, 1996.

"Saving Our Good Earth: A Call to Action." Barataria-Terrebonne Estuarine System Characterization Report. Thibodaux, LA: Barataria-Terrebonne National Estuary Program, NSU Campus, 1995.

"Saving Our Good Earth: The Barataria-Terrebonne Estuary Partnership Takes Action." Barataria-Terrebonne National Estuary Program. Stakeholders Report, 1999. Thibodaux, LA: Barataria–Terrebonne National Estuary Program, NSU Campus, 1999.

Chapter 12

Levy, Walter, and Christopher Hallowell. *Green Perspectives: Thinking and Writing About Nature and the Environment*. New York: HarperCollins, 1994.

Marsh, George Perkins. *Man and Nature*. Cambridge: Harvard University Press, 1973 edition.

Muir, John. *Our National Parks*. New York: Houghton Mifflin and Company, 1902.

Pinchot, Gifford. *The Fight for Conservation*. Garden City, NY: Doubleday, 1910.

Suggested Additional Reading

Barry, John M. *Rising Tide*. New York: Simon & Schuster, 1997.

Berry, Wendell. *The Unsettling of America: Culture and Agriculture*. San Francisco: Sierra Club Books, 1977.

Dean, Cornelia. *Against the Tide*. New York: Columbia University Press, 1999.

Dobbs, David. *The Great Gulf: Fishermen, Scientists, and the Struggle to Revive the World's Greatest Fishery*. Washington, D.C.: Island Press, 2000.

Donahue, Brian. *Reclaiming the Commons*. New Haven: Yale University Press, 2000.

Ehrenfeld, David. *The Arrogance of Humanism*. New York: Oxford University Press, 1978.

Gore, Al. *Earth in the Balance*. New York: Houghton Mifflin, 1992.

Hallowell, Christopher. *People of the Bayou*. New York: E.P. Dutton, 1979.

Hertsgaard, Mark. *Earth Odyssey*. New York: Broadway Books, 1998.

Kane, Harnett T. *Louisiana Hayride: The American Rehearsal for Dictatorship, 1928-1940*. New York: William Morrow & Co., 1941.

Kurlansky, Mark. *Cod*. New York: Walker Publishing, 1997.

Leopold, Aldo. *A Sand County Almanac and Sketches Here and Now*. New York: Oxford University Press, 1949.

Levy, Walter and Christopher Hallowell. *Green Perspectives: Thinking and Writing About the Environment*. New York: HarperCollins, 1994.

Liebling, A.J. *The Earl of Louisiana*. Baton Rouge, LA: Louisiana State University Press, 1970.

McKibben, Bill. *The End of Nature*. New York: Random House, 1989.

McPhee, John. *The Control of Nature*. New York: The Noonday Press, 1989.

Nash, Roderick. *The American Environment: Readings in the History of Conservation*. Reading, PA: Addison-Wesley Publishing Company, 1968.

Safina, Carl. *Song for the Blue Ocean: Encounters Along the World's Coasts and Beneath the Seas*. New York: Henry Holt and Company, 1998.

Terborgh, John. *Requiem for Nature*. Washington, D.C.: Island Press/Shearwater, 2000.

Williams, T. Harry. *Huey Long*. New York: Alfred A. Knopf, 1969.

Index